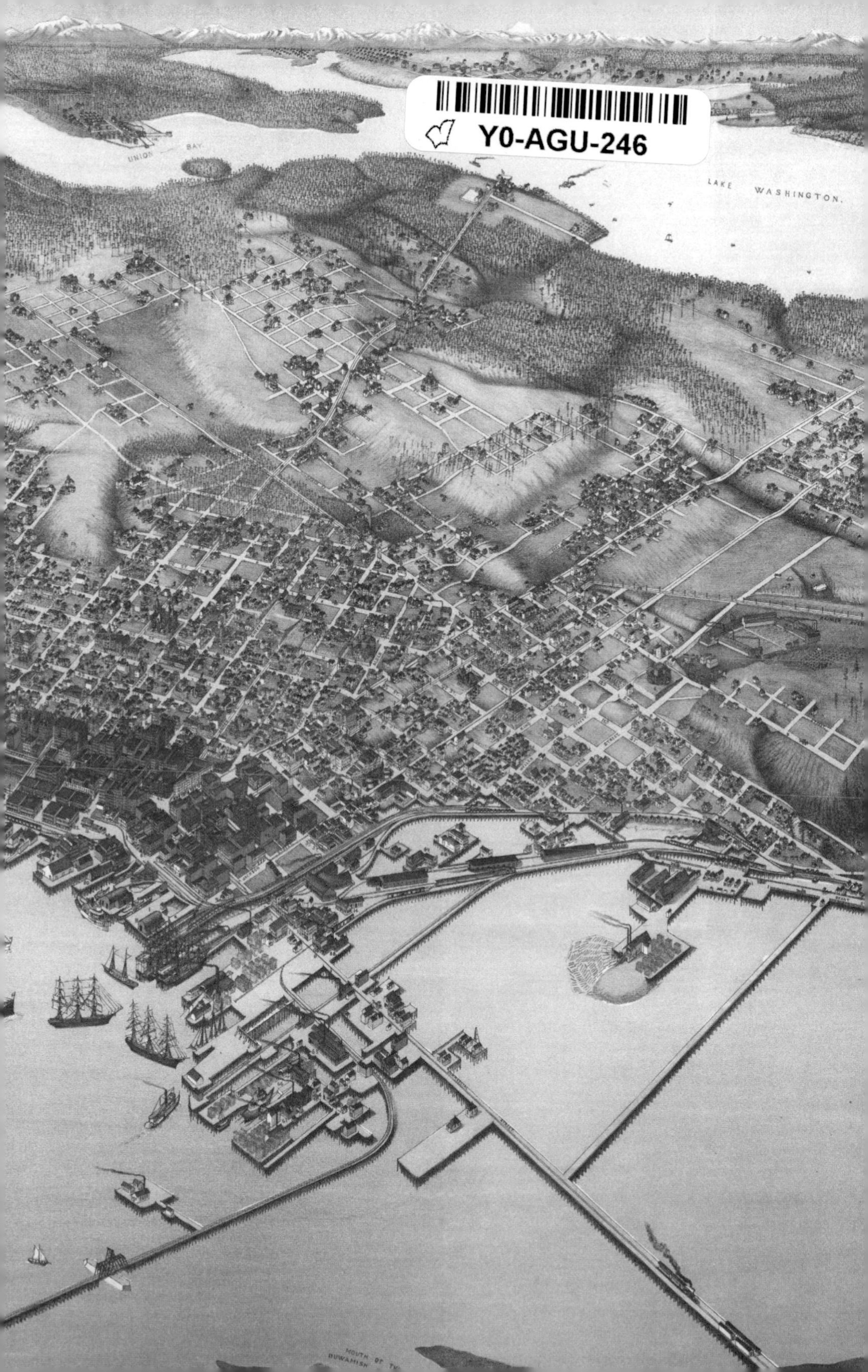
UNION BAY
LAKE WASHINGTON

TOO HIGH & TOO STEEP

To INA,
With pleasure,
[illegible signature]

David B. Williams

TOO HIGH &

Reshaping Seattle's Topography

UNIVERSITY OF WASHINGTON PRESS *Seattle and London*

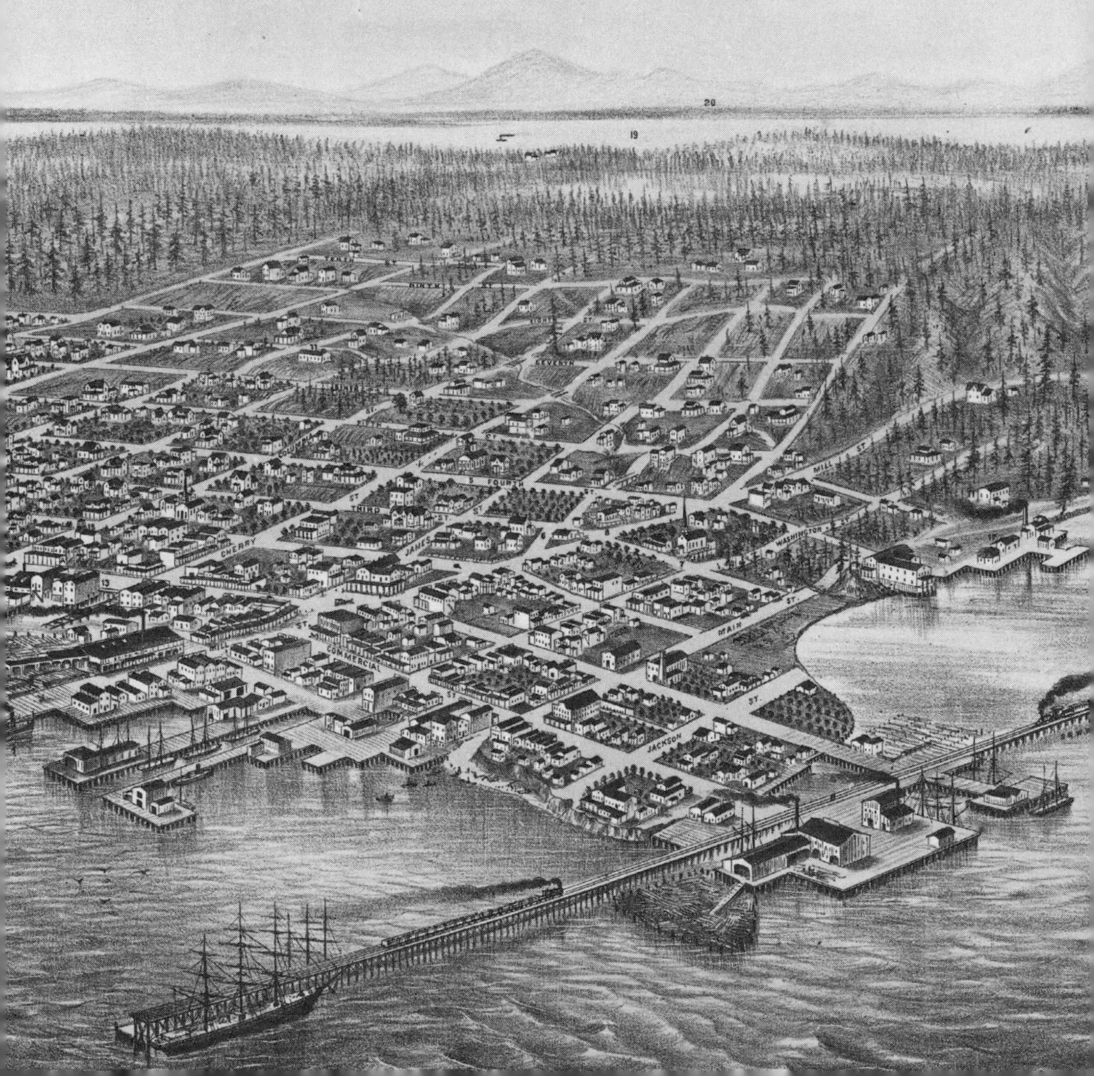

Too High and Too Steep was published with the support of the NORTHWEST WRITERS FUND, which promotes the work of some of the region's most talented nonfiction writers and was established through generous gifts from Linda and Peter Capell, Janet and John Creighton, Ruth and Alvin Eller, James A. and Katherine M. Nelson, Michael J. Repass, Linda Chalker-Scott and Jim Scott, Charyl Kay and Earl Sedlik, Robert Wack, Jacqueline B. Williams, and other donors.

4 CULTURE KING COUNTY LODGING TAX

This publication was also supported by a grant from the 4Culture Heritage Special Projects program.

Printed and bound in the U.S.A.
Design by Thomas Eykemans
Composed in Chaparral,
typeface designed by Carol Twombly
Display type set in Duke,
designed by James T. Edmondson
18 17 16 15 5 4 3 2 1

UNIVERSITY OF WASHINGTON PRESS
www.washington.edu/uwpress

ENDPAPERS: (*front*) Augustus Koch, *1891 Birds Eye View of Seattle*, Library of Congress, Geography and Map Division. (*back*) B. Dudley Stuart, *Seattle's Coming Retail and Apartment-House District*, courtesy University of Washington Special Collections, G 4284 S4:2D46 A3 1917 S72.

TITLE PAGES: Eli S. Glover's *Bird's-Eye View of the City of Seattle*, 1878 (detail).

EPIGRAPHS: Enclosure to a letter from Reginald H. Thomson to William McLeod Raine, July 31, 1911, Reginald H. Thomson Papers, Box 4, Book 10, Letter 549, Accession 0089–001, University of Washington Special Collections; George S. Turnbull, "Moving Mountains into the Sea," *The Northwest* 1, no. 1 (May 1907): 7–14.

LIBRARY OF CONGRESS
CATALOGING-IN-PUBLICATION DATA
Williams, David B., 1965–.
Too high and too steep : reshaping Seattle's topography / David B. Williams.
pages. cm.
Includes bibliographical references and index.
ISBN 978-0-295-99504-5
(hardcover : alk. paper)
1. Geotechnical engineering—Washington (State)—Seattle—History. 2. Earthwork—Washington (State)—Seattle—History.
3. Geology—Washington (State) I. Title.
TA705.3.W2W55 2015
624.1'5109797772—dc23
2015013226

The paper used in this publication is acid-free and meets the minimum requirements of American National Standard for Information Sciences—Permanence of Paper for Printed Library Materials, ANSI Z39.48–1984. ∞

For my mom, whose inspiration and support have been boundless and whose passion for history set me on the path to write this book

Seattle is builded on a group of hills lying between Puget Sound and Lake Washington. These hills are said to be glacial moraines; any way they do not lie in any orderly shape or manner; some of them stretch northerly and southerly, others easterly and westerly. These east and west hills are the ones which have given Mr. Thomson the most trouble. He says that nature intended the main traffic of the Pacific Coast to move northerly and southerly, and so when a glacier dropped a moraine across that course, that some one was sleeping and that the labor of removing the obstruction had been thrust upon us.

—Reginald Heber Thomson, city engineer, July 31, 1911

Denny Hill . . . has been marked for slaughter. . . . [The] beautiful face of the country . . . [has] been disfigured to suit the inexorable demands of commerce. . . . These hills were placed here by nature; it was never intended that they should be marred. . . . That was the old spirit. Times have changed. . . . Hills that would be permanent in other places are attacked and removed[,] . . . the steam shovel, dauntless, inexorable, beating tirelessly, against the massive side of the hill. . . .

The dirt hill . . . is doomed. . . . [A] gunner training his piece on a hostile fort . . . attack[s] the most vulnerable point of the enemy. . . . Night and day the ceaseless fight proceeds. It is a one-sided contest. . . . It is for progress; it is an outgrowth, a characteristic expression, of the Seattle Spirit.

—George S. Turnbull, *The Northwest*, May 1907

Contents

Illustrations

FIGURES

MAPS

Preface

When I was a child, my father worked with Brewster Denny, great-grandson of one of Seattle's founding fathers, Arthur Denny. I thought this was pretty neat, and I remember interviewing Brewster in third grade for an assignment I had on Seattle history. I don't know if it was during this interview or later in life, but at some point I learned that Brewster had known Arthur's son Rolland, who was six weeks old when the schooner *Exact* dropped off the twenty-two people considered to be Seattle's founding families. I was amazed that I knew someone who knew someone who had actually arrived on the boat that started the story of Seattle. It led to my lifelong enthusiasm for Seattle and its stories. That connection to the Denny family also cemented the idea that I live in a place with a relatively short history that is both accessible and tangible.

As I got older, I continued to be fascinated with Seattle's past, and I tried to weave that passion into a previous book, *The Seattle Street-Smart Naturalist: Field Notes from the City*. My goal was to explore the intersection between people and the natural world in what many call the least wild place on earth, the urban environment, while at the same time considering Seattle's historical connections to this theme. One of my favorite chapters was about the seven hills of Seattle: how they formed during the last ice age, and how they have influenced our actions in the modern city.

Too High and Too Steep: Reshaping Seattle's Topography is an outgrowth of my seven-hills chapter; in it I tell the story of Seattle through its topography. Those of us who grew up in Seattle with duck-and-cover earthquake drills and the eruption of Mount Saint Helens learned more viscerally than children in more stable locations that geologic forces played, and still play, a significant role in shaping the landscape. What we didn't always learn is the extent to which humans have been a critical force in altering the environment. By making changes ranging from lopping off hills to covering over tideflats to reengineering lakes, Seattleites have created a surprisingly unnatural landscape that affected city residents in the past and still affects us every day, often in ways we do not perceive. This is the subject of my book.

Despite all the changes to the landscape, there is little evidence obvious to the casual observer that shows how we have reshaped the topography. With some sleuthing, though, stories can be unearthed that reveal the changes our predecessors made. The stories can be seen in parking lots where cars sit catawampus to each other because the ground underneath is a former tidal swamp, and in train tracks that make a broad sweep instead of going straight because the tracks were laid before the lowering of Lake Washington. The stories can be found in borehole cores filled with the detritus of bygone industries and in early newspaper accounts of the drama of a young town determined to be a big city.

The existence of these clues inspired me to write *Too High and Too Steep*. Throughout the book I point them out in, I hope, a manner that encourages you to get outside and explore Seattle's natural and unnatural landscape. Wherever possible I have tried to bring you with me as I describe getting out on the ground, looking at the buildings, and probing the subsurface. When that hasn't been possible, I have turned to photographs and maps in order to show you features that no longer exist or that no longer look like they did before and during landscape-altering projects. My hope is that you will come away with a better understanding of not only how we have shaped the landscape but also how the landscape has shaped Seattle and those who live here.

Acknowledgments

As always happens when I write a book, I ask questions, lots of them. I thank the following individuals for responding cheerfully and helpfully to my pestering.

Scientists and researchers in various fields: Thank you for helping me better understand your research and for sharing your passion for your subjects. These individuals include Sally Abella, George Blomberg, Brian Bodenbach, Derek Booth, Brian Collins, Kurt Fresh, Sarah Gage, Andrew Gendaszek, David Giblin, Mark Holmes, Arthur Lee Jacobson, Bill Laprade, Estella Leopold, Katherine Maslenikov, Tom Pierson, Pat Pringle, Brandy Rinck, James Rufo-Hill, Sam Safran, Brian Sherrod, Hugh Shipman, Si Simenstad, Jason Toft, Kathy Troost, Jim Vallance, Glenn VanBlaricom, and Lara Whitely Binder.

Historians and archaeologists: By helping me in ways that ranged from sharing maps to answering my minutiae-filled questions, you brought your areas of research to life and helped me more fully get to know Seattle, its history, and its connection to the greater world. Among these individuals are Steve Archer, Kurt Armbruster, Kevin Bartoy, Paula Becker, Skip Berger, Matthew Booker, Sherry Boswell, Valerie Bunn, Chris Carlson, John Chaney, Susan Connole, Dotty Decoster, Paul Dorpat, Lisa Ruth Elliott, Maureen Elenga, Kim England, Nancy Faegenburg, Diana James, Paula Johnson, Rob Ketcherside, Roger Kiers, Matt Klingle, Warren King George, Kate Krafft, Larry Kreisman, Greg Lange, Peter Lape, Priscilla Long, Nathan Masters, Lorraine McConaghy, Michael Mjelde, Kit Oldham, Margaret O'Mara, Jennifer Ott, Laura Phillips, Nancy Seasholes, Stacy Schneyder, Mimi Sheridan, Charlie Sundberg, Paul Talbert, Thaisa Way, Robert Weaver, and Robert Webb.

Thanks to those of you who helped in a variety of ways, from discussions over coffee to providing me with your library card number to taking me on tours to answering my oddball questions. Each of you has helped push this book along: Ron Aldridge, Carol Arnold, Tom Berry, Cathy Britt, Curt Brownfield, Greg Brownfield, Vivian Chan, Jessica Czajkowski, Beth Deweese, Ron Edge, Lynn Ferguson, Howie Frumkin, Joseph Gellings, John Goff, Eric Hanson, Dawn Hemminger, Laurie Hutchinson, Thomas Kelly, Pamela Long, Doug MacDonald, Steve Mague, Paula McCarthy, Faithe McCreery, Laura Newborn, Brandon Oyer, Peter Paravalos, Nelson Salisbury, Joe Starstead, Liz Stenning, Coll Thrush, Katie Trujillo, Mikala Woodward, and Frank Zoretich.

Extraordinary librarians and archivists: Your assistance is the lifeblood of researchers. Every writer should be so fortunate as to work with outstanding people like the ones I have gotten to know. Thank you. Thank you. Thank you. They include Trevor Bond (Washington State University), Nicolette Bromberg (UW Special Collections), Megan Carlisle (Eastside Heritage Center), Scott Cline (City of Seattle Archives), Scott Daniels (Oregon Historical Society), Jodee Fenton (Seattle Public Library), Ann Ferguson (Seattle Public Library), Anne Frantilla (City of Seattle Archives), Sarah Frederick (Eastside Heritage Center), Mary Hammer (Washington State Archives), Karl House (Puget Sound Maritime Historical Society), Julie Irick (City of Seattle Archives), Anne Jensen (UW Special Collections), Bo Kinney (Seattle Public Library), Carolyn Marr (Museum of History and Industry), Patty McNamee (National Archives at Seattle), Pat Pierce (Bothell Historical Museum), Carla Rickerson (UW Special Collections), Josy Rush (City of Seattle Department of Planning and Development), George Shaner (National Archives), Elizabeth Stewart (Renton History Museum), Shelly Trulson (U.S. Army Corps of Engineers), Jordan Wong (Wing Luke Museum), and Gary Zimmerman (Pioneer Association of the State of Washington).

University of Washington Press: Thank you for your ongoing support of my book as well as for all you do for publishing in the Pacific Northwest. I thank, especially, Peter and Linda Capell, Beth Fuget, Bonita Hurd, Rachael Levay, Nicole Mitchell, Mike Repass, Natasha Varner, Jacqueline Volin, and Tim Zimmermann. I would also like to thank those who contributed to the press's Northwest Writers Fund. Your support has been a true gift.

To my fellow writers: Keep up the great work, and thanks for your support! Among these writers I count Langdon Cook, Beth Geiger, Lyanda Lynn Haupt, David Laskin, Lynda Mapes, Susan McGrath, Julie Monahan, Diane Sepanski, Michele Solis, Peter Stekel, and Lisa Wogan.

Thanks to Andy Nettell and Scott Wanek for reading the manuscript and providing helpful and supportive suggestions.

Thanks to my pals in the Unspeakables and to Ivan and Carol Doig: Your ever positive and helpful support has been essential to making this book a reality.

To the City of Seattle Office of Arts and Culture and 4Culture: Thanks for your financial support. We are so blessed in this community to have public agencies that support artists and historians.

Thanks to Jordan West Monez for your wonderful maps, and to Amir Sheikh for your assistance throughout with maps and thoughts on how to visualize Seattle.

Thanks to Regan Huff: Wow, I could not ask for a better editor and supporter. It has truly been special and wonderful to work with you. You rock!

Thanks to Marjorie Kittle: Golly, Ned, you are the best.

I couldn't have written this book without all of you. Any mistakes are certainly not your fault, however. Only I could have created them.

Time Line of Topographic Events in Seattle

17,400 years before present (B.P.) Puget lobe advances through Seattle
16,400 years B.P. Puget lobe retreats through Seattle
5,600 years B.P. Osceola lahar flows down Mount Rainier and reaches almost to Seattle
1,100 years B.P. Last movement on Seattle Fault
1851 November 13: Denny party arrives on the *Exact*
1852 February 15: First land claims filed in the city of Seattle
1856 January 26: Battle of Seattle
1872 March 27: First railroad runs in Seattle
1873 Ordinance No. 44 passes, giving tideflats to Seattle and Walla Walla Railroad and Transportation Company
1874 May 1: Picnic to launch construction of S&WW railbed
1876 June 8: Ordinance No. 112 passes, providing funds for grading Front Street
1877 March 7: Rail service on S&WW begins
1882 March 15: Ordinance No. 259 passes, granting rights to what becomes the Ram's Horn
1885 April 15: Seattle Lake Shore and Eastern Railway Company established
1887 January: Ordinances Nos. 804 and 806 pass, leading to formation of Railroad Avenue
1887 October: SLS&E begins service
1889 June 6: Great Seattle Fire
1889 November 11: Washington State established
1895 July 29: Eugene Semple begins the work of filling in tideflats
1898 March: Work begins on the first regrading of Denny Hill, on First Avenue

1901 November 15: Semple begins cutting South Canal through Beacon Hill

1903 August: Work begins on second regrading of Denny Hill, on Second Avenue

1904 May: The cutting of South Canal at Beacon Hill halts

1905 Opening of Great Northern Tunnel

1906 March 4: Second Avenue regrading project completed

1906 May: Work begins on third regrading of Denny Hill

1907 May: Work begins on Jackson Street regrading project

1907 Work begins on the filling in of Harbor Island

1908 August: Work begins on fourth regrading of Denny Hill

1909 September: Dearborn Street regrading project begins

1909 December: Jackson Street regrading project completed

1910 Harbor Island fill completed

1911 June: Fourth regrading of Denny Hill completed

1911 Construction begins on Lake Washington Ship Canal and Hiram M. Chittenden Locks

1912 October 4: Dearborn Street regrading project completed

1916 Building of first seawall, from Washington to Madison Streets

1917 July 4: Hiram M. Chittenden Locks officially open

1929 May: Work begins on fifth regrading of Denny Hill

1930 December 10: Fifth regrading of Denny Hill completed

1936 April: Extension of seawall north from Madison to Bay Streets completed

TOO HIGH & TOO STEEP

Introduction

WITHIN MONTHS OF SETTLING AT ALKI POINT IN NOVEMBER 1851, Seattle's founding families realized they had made a mistake. Their location on a forested point at the westernmost edge of the recently named Elliott Bay did not offer the good harbor necessary for the new community's success. Trade vessels from the Pacific Ocean could travel down Puget Sound to Elliott Bay, but without a well-protected deepwater port, the traders might not land and pick up the lumber the settlers hoped to sell. So in the middle of winter, Arthur Denny, William Bell, and Carson Boren set out in a canoe to sound the eastern, more protected waters of the bay.

As Bell and Boren paddled, Denny "heaved the lead"; that is, he tossed horseshoes attached to a heavy line overboard and determined the water's depth. The trio soon found a location with deep enough water on the bay's eastern edge, where wind and waves coming down the sound had less of an impact than at Alki. Paddling back to their original home, they were "not only well pleased with the excursion, but thoroughly satisfied as to the fitness of the bay as a harbor."[1]

Despite the enthusiasm of the three satisfied men, their newly chosen site was far from ideal. Steep hills and bluffs surrounded it, high tides made an island out of part of it, and to the south lay the vast tideflats of the Duwamish River, half the day an open expanse of mud, half the day covered by the water of Elliott Bay. In addition, there was a distinct lack of level land upon which to build the great city these early settlers envisioned. But the deep, glacier-derived harbor was fixed in place, and the settlers would have to adapt. With a former surveyor, Arthur Denny, as their leader, they knew the land around their future city could be changed.

Sundown over the Olympics and Puget Sound. Wikimedia Commons.

And change it they did, initially focusing on better access to Elliott Bay, the primary connection not only to the other little communities starting to pop up around Puget Sound, which could be reached only by boat, but also to the wider world, principally San Francisco and to a lesser extent Portland, Oregon. The pioneers accomplished this change by trimming the seaside bluffs and regrading the steep streets that prevented an easy connection to the waterfront.

The city's business community realized, though, that if they wanted to achieve what they perceived as Seattle's destiny of financial greatness, they also needed rail access to the rest of the country. The settlers had a problem here as well: the city had little space for a railroad. The only possible route into Seattle was across water, either below the steep bluffs north and south of the business district or across the capricious tideflats.

Unfazed by this difficulty, Seattle's early citizens decided to bridge the tidal expanse by building a two-mile-long trestle, which they accomplished by driving hundreds of wood piles into the mud. They followed the first construction with a spiderweb of additional trestles, and within a handful of years, trains crisscrossed the tideflats and also traveled north around the downtown waterfront. Not content to merely bridge these watery gaps, the Seattleites decided to fill in the tideflats and create level ground for industry and the growing rail network. By the end of the nineteenth century, Seattle and its striving men and women were well on their way to making more than twenty-two hundred acres of new land within the tideflats. Throughout the book, I use the term *made land*, or *new land*, to refer to the process of filling in tideflats. As historian Nancy Seasholes argues in her book *Gaining Ground: A History of Landmaking in Boston*, the terms *landfill* and *land reclamation* don't work in this case. *Landfill* "evokes images of garbage dumps but can also mean fill added on top of existing land." *Land reclamation* is the wrong term because it involves "diking, pumping and draining to reclaim land from the sea." I am not writing about either of these processes but about the act of consciously making new land by dumping material in areas regularly covered by water, which is what took place in Seattle.[2]

The city's belief in itself and its eventual rise to a world-class city became known as the Seattle Spirit. As one citizen put it in a letter to the editor of the *Seattle Post-Intelligencer*, "It's an armored 'spirit,' many-armed, argus-eyed, zealous, intensely loyal and ever ready and anxious to battle for, protect, build up and make of Seattle the 'metropolis of the Union.'"[3] In some ways, the Seattle Spirit was similar to the pride expressed by most young cities, but what set it apart was that Seattleites labeled it and truly seemed to believe in it. The Seattle Spirit was not simply platitudes uttered by newspaper editors, the chamber of commerce, and politicians. It was a force, perhaps best exemplified in 1874, when nearly every resident gathered on the town's outskirts to start building what in essence was a citizen-funded railroad across the Cascade mountains, intended to extend almost three hundred miles to Walla Walla, the largest city in the territory. The railroad made it only a bit past Renton, fourteen miles south; but in the eyes of those who lived in Seattle at the time, it showed that they would do whatever was necessary to make their new hometown succeed.

Seattleites also expressed this force in their relentless approach to the city's topography. Need to get from point A to point B and a hill is in the way? Lop off its top. Need better access for business? Shave away the unprofitable territory. Need flat territory for industry? Fill in the tideflats. This was the Seattle Spirit as demonstrated through steam shovel, hydraulic giant, and dynamite.

Or, as local newspaperman Welford Beaton summed up the historic viewpoint on the topography in his history of Seattle, *The City That Made Itself*: "Hills raised themselves in the paths that commerce wished to take. . . . And then Man stepped in, completed the work which Nature left undone, smoothed the barriers, and allowed commerce to pour unhampered in its natural channels."[4]

◆ ◆ ◆

What is missing from this traditional version of the Seattle creation story is that people already inhabited this land when the Denny party arrived. Native people dotted the landscape, primarily living along the various bodies of water, including Elliott Bay, Lake Washington, Lake Union, the Black River, and the Duwamish River. They called themselves People of the Inside Place, or du-AHBSH in the Native language, a name later Anglicized to *Duwamish*. Other Duwamish-related groups were known as Lake People and Shilshole. They lived in cedar-plank longhouses seasonally and year-round.

Like the Denny party, the Duwamish relied on the land's natural resources. Historian David Buerge has written that the Duwamish year began with the emergence of salmonberry shoots and fiddlehead ferns.[5] Soon roots and camas would be dug up, as nearby meadows turned green with new growth. Later, the winter villages would disperse, with groups paddling down to Elliott Bay for salmon and butter clams, and early summer would see a return to the river and the harvesting of the great salmon runs. The runs peaked in early fall, when Yakima, Klickitat, Haida, and Kwakiutl would join the locals for a time of sharing, trading, and gambling. As the big-leaf maples and red alders glowed in their fall colors, families would begin their return to the winter villages. "Life could be hard, but generally it was good," Buerge concludes, for people whose lives were so intimately connected with the landscape.

Although this book focuses primarily on the impacts of the post–Denny party occupation of the land, I do weave Native people in to the story for several reasons. They did have an impact on the land, primarily through middens, or garbage piles. The mounds often grew horizontally, in pulses, as people piled up shells, built atop them, then repeated the process, which could extend the shoreline many yards out into the water. One of the largest middens is at Discovery Park, a site people inhabited for four thousand years. Along with periodic landslides, the middens helped reshape the seaside environment from a rocky shore rich in mussels to a sandy beach ideal for clams. Other midden sites have been found at the mouth of the Duwamish River, near Shilshole, and along Elliott Bay. Although they were not on the scale of the large engineering projects I write about here, the accumulations of shells do indicate that everyone who has inhabited the area we call Seattle has had an effect on the landscape.

By the time that the Denny party arrived in Seattle, smallpox and other diseases had drastically reduced the populations of Native people, though Denny describes as many as a thousand gathering at Alki Point. More changes came with the new settlers. Native and newcomer "spoke two mutually unintelligible languages of landscape," writes Coll Thrush, in his thoughtful book *Native Seattle: Histories from the Crossing-Over Place*. "Where indigenous people saw . . . the wealth of the land as it was and had been—Denny and the others saw the wealth of the land as it could and would be, expressed in words like 'arable,' 'improvement,' and 'export.'"[6]

Although they were not explicitly hostile, the settlers expressed their differences by enacting laws and treaties that gave them ownership of the land and forced Native people off their historic territories. In this regard, not all land in Seattle was viewed as equal. Marginalized people such as the Duwamish, as well as transient settlers and Chinese workers, often ended up on marginal land, primarily the made land of the Duwamish River tideflats and what early on was referred to as Skid Road, but also Seattle's infamous Ballast Island—literally an island in Elliott Bay made of rocks jettisoned from trade vessels—and in shantytowns built on beaches. More established people lived on higher ground, away from industry, on land far less likely to suffer from tidal influences and the foul aromas of decaying waste.

Ironically, these same marginalized people were also the ones who helped change the landscape. As happened throughout the West, they were the work crews. In Seattle, they helped build the railroads, cut the first canals between Lake Union and Lake Washington, and regraded city streets. Unfortunately, scant evidence of their stories made it into historical documents, but where I can, I try to convey the impact of Seattle's landscape alteration upon the Native people and other marginalized communities.

◆ ◆ ◆

Seattle was not alone in reshaping its topography. From Boston to San Francisco, Chicago to New Orleans, landscape alteration fueled the economic engines of cities large and small. It led to better ports, safer facilities, easier rail access, and improved public health. It was done along rivers and around islands, and in lakes, estuaries, and bays. Changing the shape of the land and bodies of water was as natural to settlers, developers, and urban boosters as building houses, cutting trees, or ignoring the rights of Native peoples.

Few people before the early twentieth century considered the environmental or sociological consequences of grand projects such as building dams and canals, mining rivers and mountains, or harvesting forests. It is how the world operated; I doubt they foresaw our natural resources ever running out or even that such changes caused irreparable harm. Seattleites were simply doing what everyone else was doing.

The residents of Boston may be the first in the country to have attempted to change a city's shape. Established on a small peninsula jutting out into a deepwater harbor, the city relied on maritime trade for survival. To facilitate commerce, the Massachusetts colonial government passed a unique law in 1641 that allowed those who owned shoreline property to extend their property rights out to the low-tide line, which led to wholesale wharf building. (Laws of the time usually allowed only the right to extend one's property to the high-tide line.) This in turn resulted in wharfing out, or the process of filling in the space between the new structures with anything from tree branches to hulls of broken ships to ballast.[7]

Wharfing out was possible because the docks extended over shallow tideflats. All that was necessary to make new land there was to dump material onto the mud under the stilted structures—which of course led to docks growing longer and longer. By 1715, builders in Boston had completed the Long Wharf, extending nearly a third of a mile across tideflats into deep water, thus allowing ships to be loaded and unloaded without an intermediary smaller vessel. Soon the mudflats completely disappeared under fill, with Boston eventually making more than fifty-two hundred acres of new land along the margins of its original town site.[8]

Builders also made new land behind seawalls, breakwaters, jetties, and bulkheads. Each of these types of stone barricades created space behind which fill could be dumped. In the 1870s, for example, the Illinois Central Railroad constructed an offshore masonry wall along the Chicago waterfront of Lake Michigan. The lagoon inside the wall became an ideal spot to dump refuse, or what one report described as "large quantities and heaps of offal, stable manure, decaying vegetable . . . and various other offensive substances."[9] Subsequent fill along the Chicago lakefront came from debris generated in the Great Chicago Fire, as well as from other urban excavations. One historian estimated that more than two thousand acres of made land has been added to Chicago along the Lake Michigan waterfront.[10]

For many localities, garbage was a primary fill material—abundant, easy to acquire, and necessary to dispose of. Historian Craig Colten refers to the active use of trash in city building as a "series of outwardly progressing waste frontiers."[11] But with the rise in the nineteenth century of the miasmatic theory of disease, the idea of garbage as a means to create new land lost some of its luster. Miasmatists held that people could contract communicable diseases by breathing noxious air, or miasma, produced by landfill rich in decayed organic matter. In order to combat miasmas, many cities started to rely more on natural fill such as sand, gravel, and rocks, though they certainly did not stop using garbage altogether as a source material. Natural fill augmented by trash was commonly used in San Francisco, which has some of the most altered shoreline in the country.

Tideflat filling began in San Francisco soon after the gold rush, when piers began to reach into Yerba Buena Cove on the city's east side. As Bostonians had done, San Franciscans did, wharfing out and then building a railroad, California's first, in order to ferry sand from the nearby dunes and dump it into the shallow cove. As the city grew out from Yerba Buena, sand remained a primary source of fill, supplemented by mud dredged from the bay; abandoned ships; rubble from the city's various fires, including the one following the 1906 earthquake; and garbage. Almost three thousand acres of tidal flats and open water were filled in this manner in San Francisco.[12]

For each city that altered its land, change was about the future. In an essay about Providence, Rhode Island, urban historian Michael Holleran writes that the city filled in a historic cove because of citizens' anxieties about Providence's place in a "nationwide hierarchy of cities." When Providence was young, it had a local outlook and was proud of the unique nature of the cove and the tides that helped clean away waste and made the waterway beautiful. As the nation changed, though, the city's businesses began to rely more on the nationwide economy. The cove gave Providence an identity but did nothing to further its economic status, so inevitably the cove had to be filled in to make room for an expanded rail system.[13]

Seattle certainly fits this description of citizens concerned about their city's place in the country's hierarchy. The desire to be a world-class city has long driven the actions of many Seattleites, from bestowing the name New York Alki—or what we now call Alki Point—on what was little more than a nanospeck of civilization in 1851 to boasting about the city's spectacular natural setting to making an ill-fated attempt at hosting the World Trade Organization in 1999. That kind of ambition is evident today in the city's elaborate plan to replace the elevated Alaskan Way Viaduct with a tunnel and grand park.

Tearing down the viaduct is Seattle's most recent and most dramatic response to its topography. Because of the city's challenging layout, wide at the north and south ends and narrow in the middle, planners have few options for moving people through the urban core. In the 1950s, the solution was the double-decker concrete viaduct that cut across the heart of the waterfront. Now, in our more enlightened era, when we value views of Elliott Bay and our connection to the waterfront, as well as worry about potential seismic issues, Seattle is spending billions of dollars on the State Route 99 tunnel to send traffic beneath the downtown hills.

Directly tied to viaduct replacement is the city's other megascale engineering project, the replacement of its historic seawall. Built between 1916 and 1936, it is being rebuilt and upgraded because of seismic concerns and a desire to improve traffic flow and make a more beautiful waterfront. Will these latest attempts to address our topographic challenges succeed in making Seattle a world-class city? We don't know, just as no one knew when making any of the previous attempts.

Seattle is clearly not alone in having an altered landscape, but it is unusual for the scale and diversity of its metamorphosis. For most cities, filling in along the waterfront was the primary means of change. Seattle followed this path, increasing the buildable acreage along the edge of its downtown business district and south across the tideflats of the Duwamish River, but it also took the unprecedented step of reducing acreage by regrading its hills. Regrading came in two forms. The first, and more common, method leveled streets that undulated by cutting down high points and filling in low ones. These were generally small-scale projects involving less than 250,000 cubic yards of dirt.

The second, and far more dramatic, regrade method leveled entire hills. The most famous was the elimination of Denny Hill at the north end of downtown. Between 1898 and 1930, Seattleites washed and scraped away more than 11 million cubic yards of Denny, reducing a double-peaked, 240-foot-high mound to a pancake-flat tabula rasa. But this was not a lone hill-leveling. An additional 6 million cubic yards of dirt and rock were removed from where Jackson Street and Dearborn Street rose up steep hills.

Many cities smoothed out their roller-coaster streets and shaved off troublesome high points; but as far as I have been able to determine, only San Francisco attempted anything on the order of Seattle. In 1869, San Francisco completed the eighty-seven-foot-deep Second Street Cut through Rincon Hill. The hope was that the new route would make the land at the south end of Rincon easier to access and thus more valuable. Costing nearly three times more than had been budgeted, the cut ended up being a complete fiasco, with Rincon Hill becoming what Robert

Louis Stevenson described as a "new slum, a place of precarious, sandy cliffs . . . solitary, ancient houses, and the butt-ends of streets." Very little hill removal was attempted after this debacle, in part because the city's first cable-car service started in late 1873.[14]

Seattleites didn't stop at the hills. They also altered the city's biggest lake when they built the Lake Washington Ship Canal and Hiram M. Chittenden Locks. Linking freshwater with salt water lowered Lake Washington by nine feet and reduced the total amount of shoreline in the city by more than thirteen miles.[15] In addition, it completely rejiggered Lake Washington's plumbing by drying up its outlet, the Black River. The effect on the economy and landscape certainly compares to that of the regrades and tideflat filling.

By 1931, at least 75 million cubic yards of material had been moved by dredging, regrading, and filling. In trying to picture the amount of soil removed, an early reporter for the *P-I* wrote, "This would mean a total of 605,000 carloads, which in 40-foot cars would reach all the way from here to Broadway, New York, and more than half way back, and at twenty miles an hour would take the great trainload ten days to pass any given point on the route." Now take this image and multiply it by six, because the *P-I* man was writing in 1909, when only 12.1 million cubic yards had been moved.[16] This is a staggering number, especially when you consider that most of it was moved with small, relatively primitive machinery. What it illustrates is the incredible drive of our predecessors, best defined as an almost mystical combination of boosterism, self-reliance, insecurity, and delusion that developed into the Seattle Spirit (please see appendix).

The Seattle Spirit still haunts the city but is less of a unifying force than it was historically. Seattleites today are somewhat more secure about our place in the world, so maybe we are driven less to prove ourselves by taking on epic land-changing projects. This does not mean we aren't status-conscious about our hometown; we express our desire for civic recognition in other ways. Seattle is also a more fragmented city now, with more diverse needs and desires that lead to one group of citizens rejecting a project favored by others. This might be described as the Seattle Spirit meets the Seattle Process. The biggest change, though, in the Seattle Spirit is our better understanding of the geological challenges of our landscape.

We are still trying to skirt the hills and valleys, still worried about the shoreline, still developing massive engineering projects. But in a fundamental way we are very different from our predecessors. Like many who headed west, the city's settlers came with the attitude that they would do whatever was necessary to make their new hometown succeed. If that meant changing the landscape, then that's what they would do, without regard to how it would affect the environment or the Native people who lived here, and often without regard to their fellow settlers. Modern Seattleites still want to improve the place where we live, but we also recognize that no project can proceed without considering social and environmental issues, as well as an entirely new set of issues including earthquakes and climate change. Whereas earlier generations of Seattleites came to change the landscape, now we worry about how the landscape will change us.

1

Geology

IN LATE 2013, I FLEW BACK TO THE PLEISTOCENE. I DIDN'T EXPECT to do so when I got on a flight to Tucson, but soon after taking off from fogbound Sea-Tac Airport, I felt like we had time-traveled back seventeen thousand years. Emerging from the white shroud into crystal blue skies, I could see a sheet of fluff spread below me. Seattle had disappeared, Bellevue too, as had Tacoma, the Kitsap Peninsula, and the islands of Puget Sound. The clouds had erased all traces of humanity. Only the peaks of the Cascade and Olympic mountains popped above the whiteness. For the first time in my life, I could imagine what the landscape must have looked like when a glacier three thousand feet thick covered the Puget Lowland.

Geologists refer to that glacier as the Puget lobe of the Cordilleran Ice Sheet, one of the two great ice masses that covered North America during what is popularly known as the last ice age, running from about thirty thousand to ten thousand years ago. Without a doubt, the tongue of ice that crept down from Canada was the most important factor in forging the landscape of Seattle.

The ice, or rather water beneath it, gave us Lake Washington, Lake Union, and Lake Sammamish. The ice left behind frozen blocks that melted into Bitter Lake, Green Lake, and Haller Lake. It carved the city's hills, ridges, and valleys; deposited the sediments that generate our landslides; and led to the formation of Puget Sound and the Duwamish River valley. Without the ice, we wouldn't have our challenging winter driving conditions, constricted traffic corridor, hard-to-dig soils, pulse-raising downtown street climbs, or stunning views. Without the ice, Seattle would be flat and boring, though much easier to navigate.

Sunrise over the Cascade mountains. Courtesy Chris Lewis.

The first big impact of the Puget lobe came when the glacier blocked the Strait of Juan de Fuca. Without a connection to the Pacific Ocean, water melting from the ice front flowed into an immense lake that spread from the Olympics to the Cascades. The ice-melt streams carried vast quantities of fine sediment, which settled in the bottom of the lake. If you have ever been to the mouth of a modern glacier and have seen milky white, sediment-choked water gushing out of the ice, you have seen this phenomenon, a by-product of the glacier's grinding the land to powder. In the Seattle area, the lake deposits form a layer up to a hundred feet thick known as the Lawton Clay.

The best place to see the clay and silt in Seattle is where it was named, Discovery Park, formerly known as Fort Lawton. Along the base of the cliff just south of West Point you can find clumps of dark-gray to blue-gray, finely layered clay and silt, which have washed down from above. It is a wonderfully smooth and dense material that looks ideal for a mud bath. It is also ideal for impressing your friends, because you can pick up what looks like a fairly solid rock and break it in half. If you do so, look at the chunks in your hands. You will see that they are densely packed, like fudge. If you pour water on the clay pieces, very little will soak in, a characteristic that has had monumental implications for landslides and the topography of Seattle.

As the ice got closer to Seattle, the sediment coming out of the glacier's snout changed. It became sandy and gravelly, evidence of deposition by high-energy streams. Known generically as advance outwash and specifically in our region as the Esperance Sand, this type of deposit is a common feature of glaciated landscapes. In Seattle, the Esperance is up to two hundred feet thick and lies directly on the Lawton Clay, a relationship clearly visible at Discovery Park. The Lawton is the dark-gray layer sitting on older, tan-to-brown interglacial sediments and is positioned under the younger, light-gray outwash beds.[1]

If you turn around and look west, out over the sound, you can see another manifestation of the layering of the Lawton Clay and Esperance Sand, or what geologist Derek Booth has labeled the "most prominent single landform of the entire region." He first appreciated this phenomenon during a meal at the Athenian Inn restaurant in the Pike Place Market.[2]

"I was gazing out the window at breakfast, when I noticed how all the hills line up. Of course, I had noticed this before, but I distinctly remember having an aha moment," he says. "At the time, I had the advantage of working in these areas: I knew what was under the hills and I knew that they couldn't have been planed down. They had to have been built up by the outwash."

What Booth observed is something probably many have noticed from any point where one has a wide view of Puget Sound, such as the Space Needle or the Bainbridge Island ferry. All the Seattle hills, the islands of Puget Sound, and the hills of the Kitsap Peninsula top out at the same elevation, four hundred to five hundred feet above sea level. In other words, just before the Puget lobe of the Cordilleran Ice Sheet arrived in Seattle 17,400 years ago, the area between the Olympics and the Cascades was covered in a level plain made of the Esperance Sand atop Lawton Clay. Only a few features, such as Green and Gold Mountains on the Kitsap Peninsula and the Newcastle Hills, Tiger Mountain, and Cougar Mountain, rose above the level plain, or what is now the tops of the hills.

Finally, the towering Puget lobe—as high as six stacked Space Needles—arrived and deposited a mantle of sand, cobbles, gravel, and boulders atop the Kansas-like plain of Esperance Sand and Lawton Clay. Best known to urban dwellers as hardpan, and to geologists as Vashon till, this mantle, the youngest of the glacial deposits, is commonly only a few to ten feet thick, with some spots as deep as thirty feet. Geologists describe it as the material "carried and 'smeared' along the sole of the glacial ice."[3] The ice then continued to move over the landscape for another hundred miles south, as far as the city of Olympia, each year covering a stretch equal in length to about five to six football fields. Ice remained at the Puget lobe's southern terminus for a time before melting, and as it melted, the terminus retreated northward. It passed back through Seattle around 16,400 years ago.

The long-term cover of ice in Seattle had three effects on the landscape. The most significant is the carving of the troughs that we call Hood Canal, Lake Sammamish, Lake Washington, and Puget Sound. The culprit, however, was not the ice, says Booth. Geologists had long hypothesized that the rasping tongue of the Puget lobe was the primary shaper of the Puget Sound topography; but starting in the early 1990s, Booth and fellow geologist Kathy Troost proposed what, to nongeologists, may be a counterintuitive idea. They suggested that rivers flowing under the ice—what geologists call subglacial water—had cut down into the relatively unconsolidated sand and clay and created those troughs.[4]

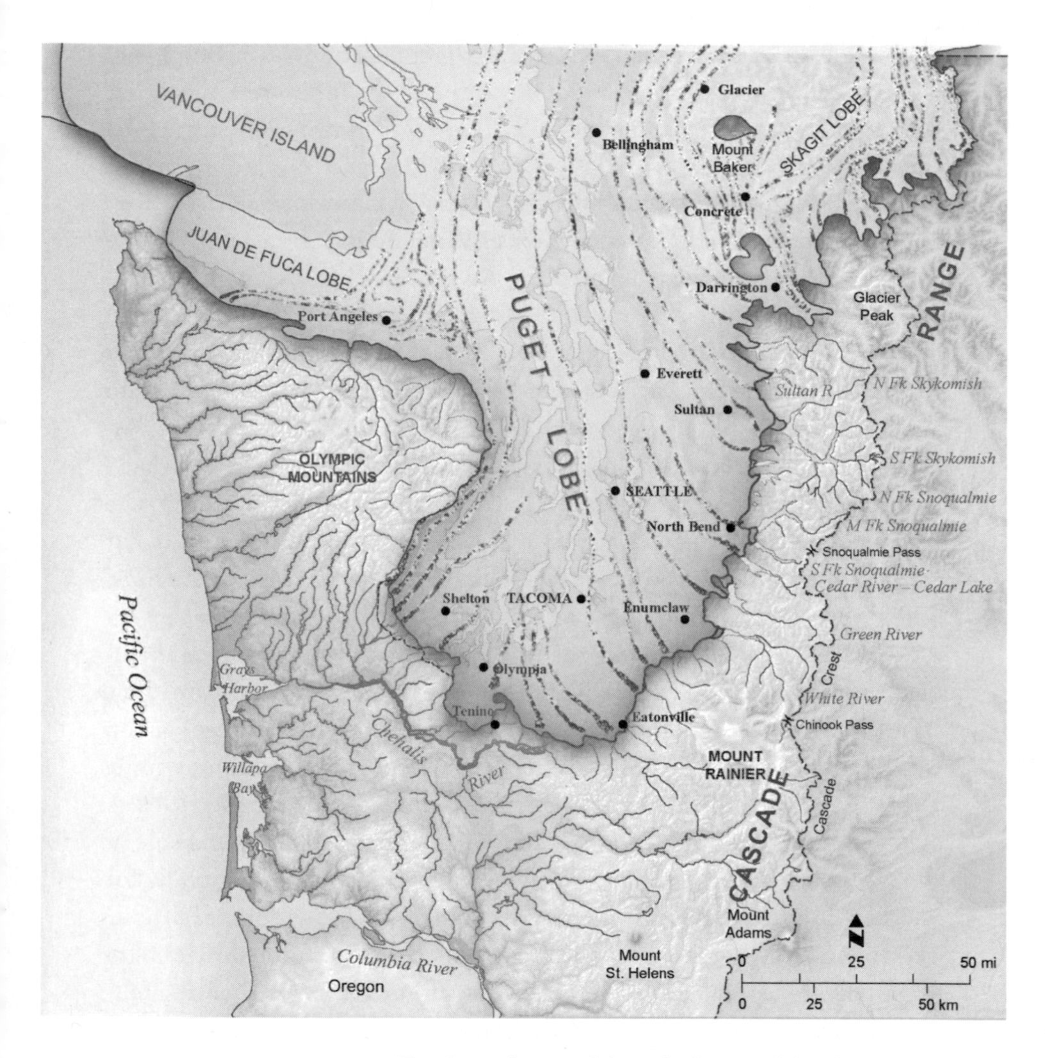

FIG. 1.1. GLACIAL EXTENT. The three-thousand-foot-thick Puget lobe glacier reached as far south as Olympia and began to melt back, or retreat, through Seattle about 16,400 years ago. The Olympic and Cascade mountains restricted its route through the area, creating hills in Seattle that trend north-south. Courtesy Washington Geological Survey, modified from Patrick Pringle, *Roadside Geology of Mount Rainier Park and Vicinity*, Washington Division of Geology and Earth Sciences Information Circular 107 (Olympia: Washington State Department of Natural Resources, 2008).

There really is no other way to carve Hood Canal, Puget Sound, and the two lakes, which have depths as great as thirteen hundred feet below the outwash plain. A look at the topography of these troughs would tell you that they are too squiggly to be cut by glacial scouring; ice is too stiff to do this, says Booth. In addition, if the ice had encountered the glacial lake in Puget Sound, the ice would have floated atop the water and had little effect on the landscape.[5]

Second, the ice acted like fingers gently running through sand, leaving behind a series of parallel valleys and ridges, or what we locals call hills. The best known of these are the famed seven that, according to local legend, Seattle was built upon. In order of highest to lowest elevation, they are Queen Anne (470 feet), Capitol (464 feet), Renton (412 feet), Beacon (364 feet), First (360 feet), Profanity (319 feet), and Denny (240 feet).[6] In recent times, Magnolia (392 feet) and West Seattle (520 feet) have replaced Renton and Profanity in the city's pantheon.[7]

The best way to experience the topographic effects of the glacier is to bike across Seattle. Traveling north or south is relatively easy as you follow one of the city's many ridges or valleys, whereas riding east or west tends to be more challenging as you ascend and descend hill and dale. In case you wondered, Capitol Hill and Queen Anne have the steepest streets, though Denny Hill also had its fair share; nearly all run, or ran, east-west.

Finally, the three-thousand-foot-thick ice sheet compressed and consolidated the underlying sediments, making them denser and harder. "Compaction completely changed the engineering characteristics of the sediments," says Bill Laprade, a geologist with the geotechnical firm Shannon and Wilson. Because the till is made of different grain sizes, it became the densest of the glacial sediments as the smaller particles filled in any voids between larger grains. In order to work it, contractors need special, toothed excavation equipment known as rippers. In contrast, contractors need only backhoes to dig the Esperance and Lawton beds. The sand is particularly easy to excavate because it lacks cohesion. The clay is more challenging because the fine-grained particles adhere to each other, making the Lawton stickier. Laprade says that the Lawton can also develop incipient cracks; and when the pressure is released, as in an excavation, large blocks can peel off, which is why contractors have to reinforce excavations for buildings.[8]

The dearth of level ground following the Puget lobe's retreat back to Canada is a landscape feature that I return to throughout the book. The region's glacial history left the would-be city with a deficit of flat land: there just wasn't much nonhilly land anywhere. In an era when large beasts supplied most of the energy to move goods on land, hills did not make a transportation-friendly city. Later, as rail began to dominate, the steep slopes and high bluffs that descended to Elliott Bay made access to the city challenging.

Although today we have left behind the era of trains and become reliant on the car, we still face many glacier-induced challenges. Cars may be more flexible in route choice than fixed rails are, and may go up hills more readily than horses, but we cannot ignore our topography. Ever try to drive across Seattle in a snowstorm or start from a dead stop on a steep hill in a manual car? Planners cannot send roads just anywhere they deem necessary for more expedient travel, which is part of the reason why the urban blight we call the Alaskan Way Viaduct was built where it was and why we are taking it down.

◆ ◆ ◆

The glacial processes I have been writing about have one thing in common. They operate at the slow pace that many define as the hallmark of geology. The next big change to hit the region was anything but slow. Known as the Osceola Mudflow, or Osceola lahar, it was a truly catastrophic event, one that caused perhaps the biggest single-day transformation of the regional landscape in recent geologic time.

It took place fifty-six hundred years ago following a volcanic explosion that blew off the summit of Mount Rainier, then more than sixteen thousand feet high. The eruption triggered a cascade of water, ice, and rock that plunged down the northeast side of the mountain at speeds in excess of 130 miles per hour. Still churning at 40 miles per hour more than twenty-five miles away, the Osceola continued to grow as it incorporated huge boulders, downed forests (which the debris had leveled), and previously deposited material in its path. Finally, about two hours and seventy-five miles later, the muddy stew slammed into what is now Puget Sound, where it had enough punch left to spread the remains of Mount Rainier for many miles underwater. During that single day, the lahar moved 4.9 billion cubic yards of material and altered more than 210 square miles, leaving deposits up to a hundred feet thick. As a comparison, the Panama Canal—often used as the measuring stick of modern earth-moving projects—took fifteen years to build and entailed moving an estimated 310 million cubic yards of dirt. Although much smaller than Osceola, the Panama Canal project still moved almost four times as much dirt as was moved in Seattle's regrade and tideflat-filling projects.[9]

Although the Osceola did not reach Seattle, it and several subsequent lahars had a profound effect on the landscape. Before the Osceola event, what we now call the Duwamish River valley was one of the deep troughs carved by subglacial water; instead of land, it was a saltwater bay that extended twenty miles south, to what is now Auburn. When the Osceola hit, the mud and muck that flowed down from Mount Rainier began to accumulate in the trough. Because the deposits were poorly consolidated, the ancestral Green River was able to erode the lahar deposits and transport more sediment into the trough. Then another lahar hit. And another and another, each dumping material into the Duwamish trough, filling it in to create a slightly sloping plain carrying the Duwamish and Green Rivers. By the time the last lahar hit, about eleven hundred years ago, the repeated flows had pushed the mouth of the Duwamish to its present location.

Around the same time, another catastrophic event hit, when a magnitude 7 earthquake thrust the ground at the Duwamish delta up by twenty feet. The quake was generated by the most recent movement on the Seattle Fault, a subsurface tear in the earth that runs east-west for twenty-five miles, from Bainbridge Island through the southern end of downtown Seattle—about where the two stadiums sit—and out east beneath Lake Washington and Lake Sammamish. Conveniently, we can blame California for this geologic problem.

For the past ten to fifteen million years, tectonic plate movements have pushed the Sierra Nevada northwest by a bit less than half an inch a year. The Sierras act like a giant block and butt into the hard mass of rock encompassing the Coast Range of Oregon, which in turn advances slightly to the north and pushes into Washington. But western Washington cannot budge, because it runs into Canada, which is part of a very stable landmass. Seattle literally sits between a rock and a hard place as its subterranean innards are squeezed by a tectonic vise.

One good place to see the long-term effects of Seattle's squeeze is a pair of anomalous mounds south of Boeing Field. You can see them best from the southbound Central Link light rail that runs to the airport. The mounds are right at eye level as you cross Interstate 5 and before you cross the Duwamish River. Look for dark-gray to black bedrock, a feature found in only a few other places in Seattle. The mound closest to the river has a nice interpretive trail to the top.

If you could slice open the 110- to 135-foot-high knolls, you would not discover any glacial material. Instead, the black bedrock consists of the oldest rocks in Seattle, the 42- to 50-million-year-old Puget Group, a several-thousand-foot-thick layer cake of many separate rock formations.[10] The rocks reached the surface because the tectonic vise has pushed the south side of the Seattle Fault up thousands of feet relative to the north side. These pimples along the Duwamish stand above their surroundings because the hard sandstone resisted glacial erosion.

What makes the hills particularly interesting is that the Puget Group rocks played a crucial role in Seattle's early economic and topographic history. Farther east, and forming parts of the Newcastle Hills and Cougar Mountain, are rich coal beds within thick layers of sandstone. The rocks of the Puget Group formed in rivers and swamps on a broad coastal plain. As younger sediments were deposited, they compacted the underlying swampy peat, slowly altering it to what would become the most valuable beds of coal on the West Coast.

The coal was initially found at or near the surface, usually by someone searching for something else. Excavating it required tunnels, or gangways, that followed the veins deep into the hills. The beds were up to twenty feet thick, but typically much thinner and often separated by thick layers of sandstone. Although the miners still needed to dig gangways up to half a mile underground, they were able to reach the coal only because it too was pushed up to the surface by the Seattle Fault; had they desired to access the coal on the north side of the fault, it would have required drilling down thousands of feet.

◆ ◆ ◆

The best place to experience the most recent effects of the Seattle Fault is near the mouth of the Duwamish River at what is known as Terminal 107 Park, not the cleverest name but a place that has become one of the most famous geologic spots in the city. From the parking lot, a dirt trail leads through shrubs down to the river, and to a bank composed of sand, silt, and clam shells. Look carefully and you will see that many clams lie on their side with the bent part of their shell pointing upward, which would have facilitated the use of their siphons for feeding and respiration.

Geologists such as Brian Sherrod of the United States Geological Survey propose that the Seattle Fault killed these clams. He notes that, in contrast to their modern location well above high tide, these clams were intertidal organisms, meaning they lived between the high and low tide lines. Something must have raised them to their present elevation. The most likely scenario is uplift during an earthquake along the Seattle Fault, which killed them and left them in their former life-position. This is the magnitude 7 quake of eleven hundred years ago, the last recorded movement of the fault.[11]

If you take archaeologists to this same spot, they will tell you a different story: The clams are not in a life position but in a death position. They are part of an extensive garbage pile, or midden, made of shells, fish, mammal, and bird bones; stone and bone tools; and charcoal and other human debris. It developed over the centuries from materials deposited by Native people who lived along the Duwamish. The archaeologists recognize that this particular bank is a bit odd because it lacks the charcoal and fire-cracked rocks usually found with middens. Nor does it have the classic black, greasy look often found in middens, but studies show obvious midden deposits in other parts of the park. The archaeologists don't dispute that the bank and the surroundings were uplifted; they just don't think the riverbank is a clear-cut illustration of uplift as do geologists.

There is one other feature to note at this location: the dark-gray sand atop the bank. Sherrod and others see it as evidence of the last of the four lahars to descend Mount Rainier. They cannot tell if the lahar, which occurred right around the same time as the last movement on the Seattle Fault, flowed all the way to this spot, or if the river carried the sediment here from a short distance away. Either scenario indicates that lahars from Mount Rainier have reached Seattle, carrying with them land- and life-changing quantities of soil and trees and rocks.

That last lahar, combined with the last movement of the Seattle Fault, had a further effect on the Seattle landscape. When the fault lifted the land and the lahar deposits by twenty feet, it must have dammed the Duwamish and created a lake behind the uplifted delta. The lake probably lasted only a handful of weeks before the river eroded through the dam, says Sherrod, but it would have taken many decades or even centuries to redistribute the uplifted lahar sediments across the delta of the Duwamish. In doing so, the river established a precedent of erosion and deposition that eventually led to the formation of the Duwamish River tideflats, one of the central geographic features in the early history of Seattle.

In addition, Ralph Haugerud, a colleague of Sherrod's at the Geological Survey, says the uplift made the Duwamish River valley better for farming.[12] Previously, the valley bottom would have had soggy soils. After rising twenty feet or so, the soils would have been much better drained and much better for growing crops, a fact initially exploited by Native people and later by the valley's other early settlers.

◆ ◆ ◆

No other site in Seattle is so exciting to Sherrod and his fellow geologists, for no other site provides such graphic evidence of the region's intertwined history of earthquakes and volcanoes. In that way, the little bank of sand and clams at Terminal 107 is sobering. It clearly shows that we live in a landscape that was shaped by and will continue to be shaped by natural forces far beyond our control. We will do our best to counter those forces with good engineering and planning, but ultimately our lives will be changed the next time Mount Rainier sends a lahar our way or the Seattle Fault shifts the ground by twenty feet.

Humbling as this location is, we can also thank the geologic forces that formed it for the development of the city. In Ralph Haugerud's words: "Seattle is here because of the Seattle Fault." Not only did the uplift lead to better soils in the Duwamish, but it also formed the landing site of the Denny party, the men, women, and children of Seattle's founding families. On November 13, 1851, the schooner *Exact* dropped twenty-two people on the little bench of land that we call Alki Point. Led by Arthur Denny, they would overwinter at Alki before eventually moving across Elliott Bay to land near what is now Pioneer Square.

That bench was and still is an unusual spot in Puget Sound. Not only did it have good wood for building, but also it was flat and above the high-tide line. Few other sites nearby had these features, and none in such an ideal combination as Alki. It was the perfect spot to land, and one long known by the Duwamish people. They called this raised bench *Sbaqʷábaqs*, or Prairie Point, an indicator of its unusual nature and location. Plant collections from the late 1800s—some of the earliest in the city—reveal that plants adapted to drier, more prairielike conditions thrived here.

Haugerud says that we can further thank the Seattle Fault for the stellar deepwater harbor, which the earliest settlers needed in order to survive. When the Puget lobe encountered the Gold and Green Mountains, Newcastle Hills, and Tiger and Cougar Mountains, the high hills constricted the glacier in a tectonic girdle, which increased the pressure on the subglacial water and made it better able to excavate downward. The result is that the deepest part of Puget Sound is just north of the Seattle Fault. Had it been shallower, sediment could have filled it in more and led to a lower-quality harbor.

Depending upon your worldview, and tolerance for risk, we are either fortunate or unfortunate that those early settlers didn't realize the reasons the location seemed ideal. "If you look at all the major population centers in Puget Sound—Bellingham, Everett, Seattle, Tacoma, and Olympia—almost all of them have big, active, ugly faults under them," says Sherrod. "We definitely would not pick those places today." Geology may have led to the formation of our city, but it also has the potential for serious destruction, a fact that drives many of our modern land-use decisions.

◆ ◆ ◆

One other type of short-duration event has long played a role in shaping Seattle's topography. To explain this phenomenon, I turn to someone who played his own prominent role in shaping Seattle.

On February 27, 1897, Seattle city engineer Reginald Thomson sent a two-page letter to the city's corporation counsel, or lead attorney. The subject was landslides, which Thomson wrote had occurred in Seattle "from a time to which the memory of man runneth not back." The reason had to do with the interaction between a layer of impervious blue clay that lay at the base of the land and pervious glacial drift atop it. When it rains, wrote Thomson, water percolates through the drift "in devious ways" until it reaches the clay below, resulting in a "condition of saturation and suspension" that continues until "the surface ground breaks and settles down."[13] Using more recent terminology, a modern geologist would describe how the pressure of water that has penetrated through

Esperance Sand increases at the point of contact with Lawton Clay and makes the slope susceptible to sliding.

Go to any steep hillside around the city and you will find evidence of landslides or future landslides. Look for areas mostly bare of shrubs or trees, such as the east side of Magnolia Hill (above the Magnolia Bridge), which slid in 1997; areas rich in springs and seeps, evidence that water has percolated down to the Lawton Clay and is following gravity to the surface; areas covered in alders and maples, trees that pioneer unstable terrain; and areas with stairs askew, which look as if they were made by a drunk contractor and indicate ground movement.[14]

The *Seattle Landslide Study* reported that more than thirteen hundred landslides had hit Seattle since 1890.[15] These included high-bluff peel-offs, groundwater blowouts, deep-seated landslides, and skin slides. Skin slides, which are small and can be triggered by intense rainfall, are the most common. Deep-seated events are the most spectacular; one of the best known in recent years took place on Perkins Lane in the Magnolia neighborhood, when part of the bluff slid and carried several houses down to Puget Sound. The landslide occurred in January, by far the most likely month for slides, with almost three times more than the next-nastiest months, February and March. Not coincidentally, November, December, and January receive the most precipitation.

If ice and water are the great sculptors of our landscape, then landslides are the fine chisels, nipping the rough edges and often leaving behind very steep, sometimes nearly vertical, slopes. Usually fairly small and localized, landslides reveal the weak spots in the landscape, the places one might not want to build. (Wave action along the city's bluffs also helped generate places that early home builders avoided, such as the nearly vertical cliffs found at Discovery Park and north of Golden Gardens Park along Elliott Bay.)

Landslides are one reason that green spaces such as Interlaken, Frink, Carkeek, and the Duwamish Greenbelt became parks and not homesites. But we have since rewritten the story line, ignoring the historic reluctance to build on clearly less-than-stable slopes. More than 80 percent of the recorded slides owe their slippage to human disturbance, including poor installation and poor maintenance of drainage systems, leaving broken pipes unrepaired, imprudent pruning of stabilizing vegetation, and the chopping off of the toes that protect slopes. This latter problem continues to exasperate hazard-prevention planners. Think of a mirror resting on a hardwood floor and leaning against a wall. One easy way to stabilize it is to put a heavy object in front of the mirror. With the weight in place, the mirror doesn't slide; without it, seven years of bad luck. The same concept applies when people remove material from the toe, or base, of a slope. Many of the landslides that have hit since 1890 would probably have occurred in geologic time; but with human activity, we have fashioned a new time scale, one based on human time. In doing so we have made landslides more relevant to the city's topography and to those who live here.

Other green spaces previously remained green because of another feature, the abundant streams that once meandered across the landscape. Nearly all have vanished under concrete or fill, but if you look at the survey maps produced by the General Land Office between 1858 and 1862, you will find dozens and dozens of waterways illustrated by the cartographers. Most started as the uncountable springs and seeps that emerged from the Esperance-Lawton contact.

The features carved by the majority of these waterways were small scale, but they also gave texture to the landscape. Where once was a smooth, rounded hill, now there were furrows and gullies, clefts and dips, and perhaps a pond or two. Growing up at the north end of Capitol Hill, my friends and I had the ravine, a playground of creeks that had carved the hillside into a scalloped landscape perfect for kids to explore. It surprised me when I later learned that the place had the formal name of Interlaken Park. As with most other locales in Seattle, none of the creeks in the ravine had names, though there are more than forty-five officially designated creeks in the city. I suspect that few of those names are known outside the neighborhood where they flow. Probably the only well-known creek names are those of the biggest: Thornton, Longfellow, Ravenna, Taylor, and Fauntleroy.

Each of these sizable streams also formed large topographic features on the landscape; some of the smaller streams created similar, but smaller, features. Part of what is intriguing about the stream-carved ravines and canyons is that some cut *across* the general grain of the glacial landscape. For example, both Ravenna and Thornton have long stretches that run northwest to southeast. Geologists do not have a good theory to explain these anomalous features. The streams could be flowing in channels incised by subglacial streams, or they could be following some unrecognized tectonic structures, which may account for Seattle's most significant northwest-southeast-trending topographic feature, the long diagonal now filled by the Lake Washington Ship Canal. It is the city's biggest geological enigma.

◆ ◆ ◆

In few other major cities has such a set of dynamic geological forces been so integral to the life of the city. Seattle is not unique in being built on a glacial landscape—Boston, New York, Chicago, and the Twin Cities are situated in such landscapes, too. But it is unusual in sitting in a basin between two mountain ranges, which caused the ice to flow and erode differently. This region's relative warmth also led to a dynamic between ice, water, and erosion that makes it differ from those locales. The result of these differences is that we have steep hills and other cities don't. Nor is Seattle unique in sitting on a geologically vulnerable landscape. All of California's major coastal cities must worry about earthquakes, and New Orleans has to plan for its subsiding waterfront and the rising sea levels. Certainly the locations of most cities have been dictated by geography, particularly by proximity to water, which is shaped by a location's geologic history; but Seattle's geology has produced a unique set of problems, from earthquakes to landslides to volcanic eruptions, that pervade our lives.

In everything from the city's founding to its earliest economic survival to its later commercial development, from where we live to where we play, from how we build infrastructure to how we disassemble the land, Seattle's fate has been inextricably tied to geology. I don't think we consciously recognize this. It is deeper than that. Only when large-scale events, such as earthquakes and volcanic eruptions, affect us do we fully admit to ourselves the forces that have shaped and continue to shape our lives in Seattle. But we also know subconsciously that we cannot escape these forces. Perhaps that is why we have so readily accepted the great changes we have wrought on the land. It is our past and it is our destiny.

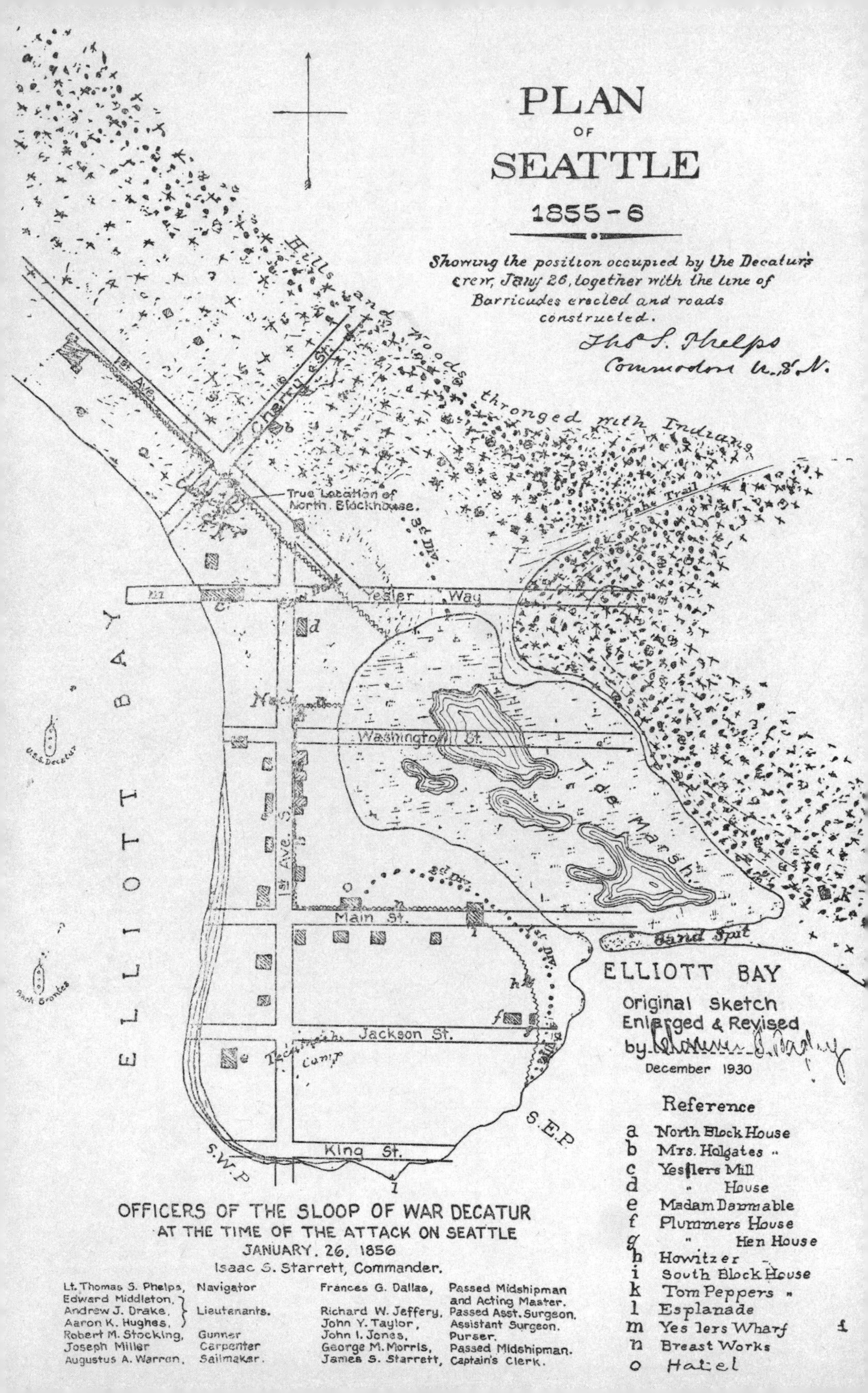
PLAN
OF
SEATTLE
1855-6
Showing the position occupied by the Decatur's crew, Jan'y 26, together with the line of Barricades erected and roads constructed.
Thos S. Phelps
Commodore U.S.N.
Hills and Woods thronged with Indians
1st Ave.
Cherry St.
True Location of North Blockhouse.
Lake Trail
3d Div.
Yesler Way
Washington St.
Tide Marsh
1st Ave. S.
2d Div.
1st Div.
Main St.
Sand Spit
ELLIOTT BAY
Original Sketch
Enlarged & Revised
by
December 1930
Jackson St.
Tecumseh Camp
King St.
S.W.P.
S.E.P.
ELLIOTT BAY
U.S.S. Decatur
Bark Brontes
Reference
a North Block House
b Mrs. Holgates "
c Yeslers Mill
d " House
e Madam Damnable
f Plummers House
g " Hen House
h Howitzer
i South Block House
k Tom Peppers "
l Esplanade
m Yes lers Wharf
n Breast Works
o Hatel
OFFICERS OF THE SLOOP OF WAR DECATUR
AT THE TIME OF THE ATTACK ON SEATTLE
JANUARY. 26. 1856
Isaac S. Starrett, Commander.
Lt. Thomas S. Phelps, Navigator
Edward Middleton, Andrew J. Drake, Aaron K. Hughes, Lieutenants.
Robert M. Stocking, Gunner
Joseph Miller Carpenter
Augustus A. Warren, Sailmaker.
Frances G. Dallas, Passed Midshipman and Acting Master.
Richard W. Jeffery, Passed Asst. Surgeon.
John Y. Taylor, Assistant Surgeon.
John I. Jones, Purser.
George M. Morris, Passed Midshipman.
James S. Starrett, Captain's Clerk.

2

Seattle's Historic Downtown Shoreline

AS MAPS GO, IT IS NOT TERRIBLY SPECTACULAR. NO GILT MARGIN, terra incognita, or fanciful sea monsters. The cartographer has included two ships, but they look more like pea pods than anything seaworthy. Drawn in pen on paper, the map depicts less than a square mile, with land making up about half the map, water about a third, and the title, legend, and white space the rest. Thomas Phelps, in service on the navy sloop of war U.S.S. *Decatur*—one of the pea pods—drew the map to illustrate the events of January 26, 1856, Seattle, Washington Territory.[1]

A little more than four years earlier, settler families had established the community they named in honor of a leader of the local Native American tribe. When Phelps arrived, he estimated that around fifty people lived in Seattle, with perhaps up to 175 total within thirty miles. The *Decatur* had sailed into Elliott Bay because of concerns about attacks from the Native people, who were being displaced by the new arrivals. Phelps had been in the area for about three months when the event depicted on the map took place. Perhaps justifying the need for the *Decatur*, Phelps noted on the map that the woods surrounding Seattle were "thronged with Indians."

Within the dab of habitation, he drew thirty buildings, including several homes, two blockhouses, two hotels, and a henhouse. On the shoreline are a wharf and a long, narrow build-

FIG. 2.1. PLAN OF SEATTLE, 1855–56. Historian Clarence Bagley updated and modified the original Phelps map in 1930. Unlike the original, Bagley's edition shows the modern streets with their names, which makes it easier to compare with present-day Seattle. Courtesy University of Washington Special Collections, UW 4101.

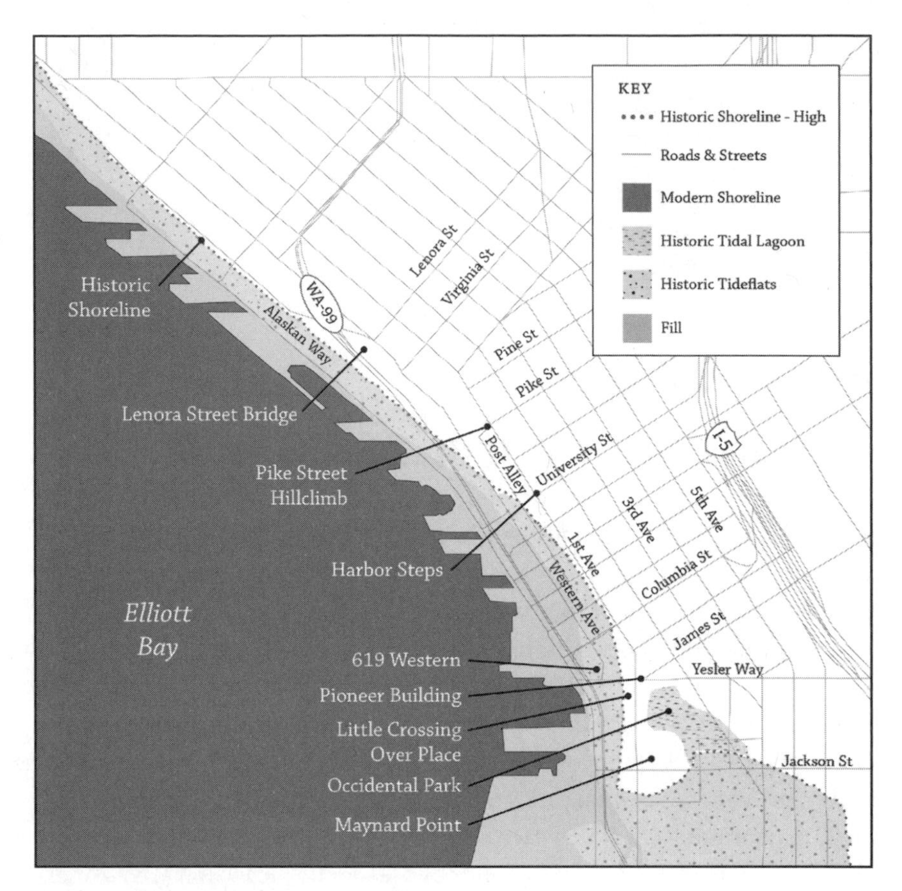

MAP 2.1. HISTORIC SHORELINE. Historical mapping by UW's Puget Sound River History Project.

ing, Henry Yesler's mill. Yesler's was Seattle's first start-up business and largest employer. Like many more modern start-ups, Yesler employed a mix of locals, or Native people, and newcomers, or settlers. Behind the mill is flat land that the new Seattleites called the "Sag," or the "Sawdust." It was where one of Yesler's employees, Dutch Ned, dumped wheelbarrow after wheelbarrow of sawdust from the mill, and where Yesler built a small house. South of the Sawdust, an isthmus called the "Neck" connects to the broad side of a lemon-shaped peninsula, where the majority of the town's buildings stand.

Carefully drawn in the center of the peninsula is a grid of streets with just one or two buildings. The area looks a bit like a pitch for a planned community. A lone building lurks on the southwestern edge of the street grid. Phelps has labeled it as Madam Damnable's, a reference to a boardinghouse that he later wrote was run by "a terrible woman, and a terror to our people, who found her tongue more to be dreaded than the entire Indian army recently encamped in our front."[2] Reputation holds that the building was thronged with women of ill repute.

Madam Damnable's is the southernmost building of what was, at the time, Seattle's longest street, which ran about six hundred feet. No other buildings on the street merited a name. On the east side of the peninsula, Phelps has drawn a swamp, or salt marsh. It's about the size of the peninsula and is formed in an area protected from wind and wave action by a sandspit that sticks out from the eastern hills and leaves only a tiny gap between the spit and the peninsula.

The map's newest features are breastworks, or a log fence, that surround the community. They had been erected by the *Decatur*'s men and the settlers to protect Seattle. On the day illustrated by Phelps's map, a warning had arrived that an "overwhelming number" of Indians, "like so many demons," waited in the woods above town.[3] Sailors and marines made up the bulk of the nonnative fighters, with most of the women and children and some men hiding in the blockhouse. After the initial shots, including mortars lobbed by the *Decatur*, the two sides battled until about 10 P.M. Two settlers and an unknown number of Natives died before the Natives retreated into the forest. The Battle of Seattle was over.[4]

I cannot get this singular map out of my head. It shows my hometown in a manner unlike any I know or even imagined. The map is the first that shows Seattle, not just the land that would become the city, but also the layout of the infant village—what would become the area that modern inhabitants know as Pioneer Square and First, Occidental, and Second Avenues, and Main and Jackson Streets. Like most people, I am used to the modern map of my hometown and its infrastructure. I see the roads that I regularly bike and drive, the home I now live in and the one I grew up in, the bridges that carry me over lakes and canals, and the parks where I run and walk. I see restaurants, theaters, bookstores, and places I have worked. I see where we spread my father's ashes, where I kissed my first girlfriend, and where I got in my first car accident.

Each of these layers forms the map of my home. It is a multidimensional picture made not simply of the visual aspects of the urban infrastructure but also of memories, aromas, sounds, and emotions. I neither see nor can imagine any of this in Phelps's map. When I look at the scattering of buildings in that tiny dot of Seattle, it makes me think of seeds in a garden. And as with a seed, there was no guarantee that it would grow and thrive. We know from writings by the Denny party that they thought it was their destiny to establish a town that would develop into greatness, but how many other specks of humanity depicted on large maps, and driven by urban optimism, had the same aspirations?

In 1856, there would have been little reason to think Seattle would survive. An earlier cartographer had labeled the lands above what would become the settlement as "Thickly Timberd."[5] Dominated by towering conifers and an impenetrable understory, the timbered land stretched for hundreds of miles to the north, south, and east, making one of the greatest and densest forests in North America. The nearest city of any size, Portland, with around twenty-five hundred people, was eight days away by schooner. The one positive aspect of Seattle was that it was located on the water and had a good harbor for shipping the great timber that engulfed it.

Phelps's map illustrates another threat to Seattle's long-term survival. What Phelps cannot show, but what early Seattleites describe, is that the peninsula where most of them lived became an island periodically. At very high tides, water would cover the Neck and completely isolate the high ground of the peninsula from the mainland. But the settlers' optimism and drive overcame their topographical and hydrological challenges. They covered up, cut down, and filled in the undesirable, and unprofitable, parts of the city, creating a landscape that would be good for business, and a landscape where the past was little present. In so doing, they also started to erase the signs of those who had lived here for millennia and who held a very different view of the place. To the Native people, the land provided an abundance of food and resources for shelter, as well as good transportation routes.

Seattle's earliest citizens brought a far different set of expectations and desires. The city they envisioned was one of straight and level streets, large homes filled with nonlocal luxuries, abundant industries, and trade with a worldwide network. All of which would require change. They were so successful that little remains of the landscape illustrated by Phelps. But it is usually impossible to erase the past; and in Seattle's case, pentimenti of the city's original shoreline can still be found.

◆ ◆ ◆

Modern Seattle has long been described as having an hourglass shape, bounded on the east and west, respectively, by Lake Washington and Puget Sound. In the neck of the hourglass sits the downtown business district, with the city's industrial base spreading to the south. Surrounding these areas are residential neighborhoods covering the famed seven hills and many other ridges and valleys. Flowing north and into Elliott Bay, which forms the western border of downtown, is Seattle's lone large river, the Duwamish.

In order to find the traces of Elliott Bay's original shoreline, I am standing about a mile north of the long-gone peninsula and marsh on the last segment of what some dubbed the Bridge to Nowhere, now known as the Lenora Street Bridge.[6] I am about forty-five feet above Alaskan Way, a multilane road that runs the length of the city's waterfront, and I'm between fifty-five and seventy-five feet above sea level. I am also sandwiched between two buildings—a hotel and a condo complex. Each rises above me, but I can still look west, toward a redeveloped pier, Elliott Bay, West Seattle, Bainbridge Island, and the Olympic Mountains. It is, as the hotel website has informed me, a *gorgeous view,* although it helps to wear my metaphorical blinders, which allow me to filter out the concrete, traffic, and gargantuan cruise ships that mar the beauty of the water, mountains, and greenery.

My interest lies in a mundane slope of gray sediments and sporadic vegetation that I claim provides one of the best insights into our city's topographic history. Nowhere else in the downtown area can you see a hillside such as this one. It is the lone reminder of what the earliest maps of the area show, the steep bluffs of sand, silt, and gravel that once edged many sections of the city's downtown.

As I stand on the Lenora Street Bridge, I imagine the scene in 1852, when Arthur Denny, Carson Boren, and William Bell paddled their canoe in search of a good harbor. A few hundred yards north of here, at the present-day Bell Street, the men had seen a ravine that cut back into a bluff. They stopped; Denny got out, clambered up the hillside, and found a gentle slope where a fire had burned the woods. He liked the spot so much that he later filed a land claim for it, with Bell filing for a site to the north and Boren a site to the south. Later, though, Denny would have regrets and write in his reminiscences that the terrain was "so rough and broken as to render it almost uninhabitable." He would continue to own the land but eventually moved closer to where the majority of Seattle residents were settling, because he found communication with Elliott Bay "next to impossible."[7]

The bluff Denny had ascended rose straight up about forty feet as a sandy face before sloping steeply back toward what would eventually be known as Denny Hill. Scattered wildflowers and a shrub or two would have grown on the exposed face, with Sitka spruce, madrona, and Douglas fir thriving on the slope. At the base of the slope sat large blocks of collapsed bluff, as well as a five- to fifteen-foot-wide shelf of sand dotted with trees and logs washed down from above and in from the sea during big storms. As on modern Puget Sound beaches, or at least the relatively intact ones, a shallow slope of sand, pebbles, and cobbles extended eighty to a hundred feet into the water from the shelf. The top of that slope was equivalent to the extreme high-tide line, about eighteen feet above the extreme low-tide line. At very low tides, most of the beach would have been exposed, revealing a slightly sloped, sandy area that ran west before descending steeply and deeply into the bay. It was this combination of beach and deep water that helped make this side of Elliott Bay so appealing to the early settlers.

The bluff was not a stable slope; waves and currents would have continuously undermined the face. Erosion on modern-day Puget Sound bluffs ranges from barely noticeable to more than two feet per year, though landslides can remove yards in a single event. That lack of stability was a good thing, because the sediment and plant debris that came off the bluff helped build up beaches and generated essential nutrients for plants and animals.[8]

One good way to envision this full scene is to go to West Point in Discovery Park. Not only are its steep bluffs similar to what Denny would have encountered above the ravine but so is the beachscape, with its shelf of sand and gravel and flat sandy beach exposed at low tide. If you are there at low tide, you can also get an insight into the ambition, or folly, of our fair city's nineteenth-century developers. Walk out on the exposed sand flats and you are walking on the Seattle Tide Lands Plat, an ambitious 1895 plan to designate and delineate more than thirty-two hundred acres of made land stretching from the mouth of the Duwamish River to Shilshole Bay. On the West Point section, planners designated more than seventy blocks, hundreds of lots, and many named streets, none of which ever came to fruition.

◆ ◆ ◆

As was typical of most of his contemporaries, Denny failed to mention that others already inhabited the land that he and Bell claimed. The area above the Bell Street ravine was known as *Babáqʷab*, or Little Prairie, in reference to open spaces in the forest south and east of the modern-day Seattle Center. Babáqʷab was a Duwamish village site with two longhouses, each about the size of a basketball court. Native residents used the site seasonally for food harvesting, particularly salal berries, and as a winter village. They also would have caught salmon along the shoreline, taken advantage of nearby freshwater springs, and used a trail that cut over to Lake Union.[9]

By the 1870s, with the continued growth of Seattle's population (about eleven hundred people were counted in the 1870 census) and the subsequent chopping down of the forest and northward spread of settler's houses, Babáqʷab changed from a formal seasonal village to more of a refuge for displaced people, most of whom moved down to the beach. Many Native people ended up living there, owing in part to Seattle's Ordinance No. 5, which stated that "no Indian or Indians shall be permitted to reside, or locate their residences on any street, highway, lane, or alley or any vacant lot in the town of Seattle." The closest point where they could reside was the north side of Bell's land claim, or north of any land already claimed by the new settlers.[10]

It is unclear how long the longhouses remained at this site; but by the late 1880s, tents, lean-tos, and cabins dotted the shoreline. Later, seasonal use resumed to some extent, when Native people from as far north as Alaska camped here on their journeys to work the hop fields in the White River valley of what is modern-day Kent and Auburn.

◆ ◆ ◆

Wanting to follow the old shoreline and to get up close to what remains of the historic bluff that had scared off Denny, I descend the west end of the Lenora Street Bridge, continue west to the Elliott Bay Trail along Alaskan Way, turn left, and walk south till I reach the end of the condo building on Pine Street. Another left and I am at the base of the bluff and a flight of steps. Just north of the stairs sit a row of triple-stacked concrete barriers, each two feet wide, two feet tall and eight feet long.

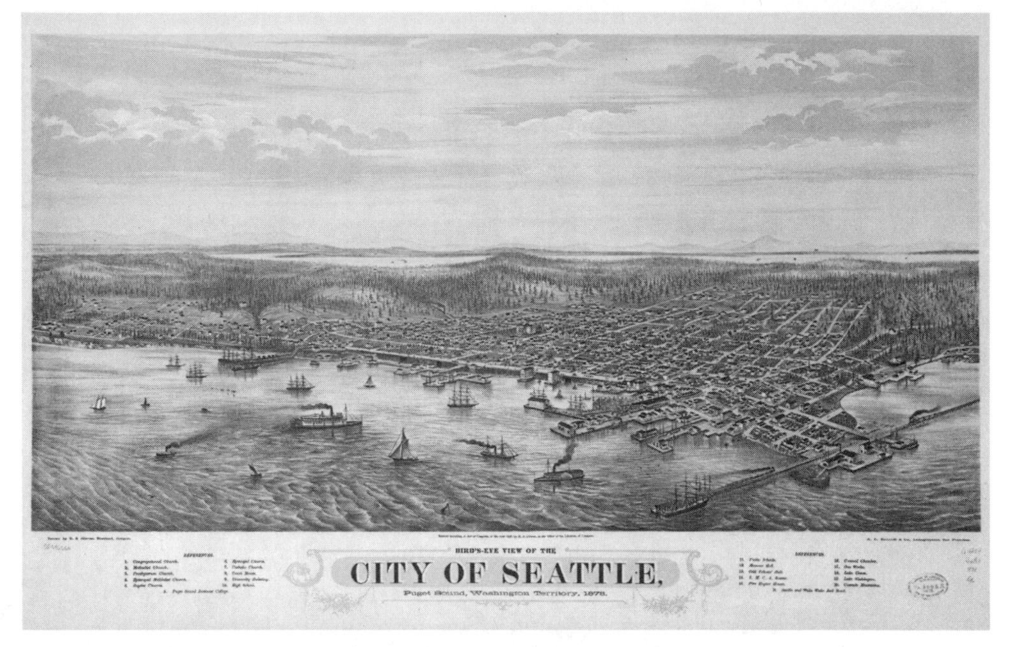

FIG. 2.2A. ELI S. GLOVER'S *BIRD'S-EYE VIEW OF THE CITY OF SEATTLE*, 1878. This is the first bird's-eye view of Seattle. In order to create his perspective, Glover had to flatten Seattle's steep hills, though Denny sort of rises at the north end of the image. The city's reliance on shipping can be seen in the abundant water traffic, but trains are just starting to edge into the story, as shown by a smoke-emitting coal train heading south from Lake Union and by two more trains converging at what is now about First Avenue South and King Street. Library of Congress, Geography and Map Division.

They illustrate the way engineers use barriers at the toe of a hill in an attempt at foiling nature. By holding back the base, they hope to prevent slides from above.

Walking south past the slope, I pass two buildings and reach the Pike Street Hillclimb, the long series of steps that connects the waterfront to the Pike Place Market. As I intersect the historic shoreline on my ascent of the hillclimb, I like to imagine that I am re-creating what would have been an ascent of the bluff that once rose here. I could also do the same thing at the ends of the next few streets south—Union, University, and Seneca—that is, climb stairs up from the waterfront to First Street. None of these streets go through, because the bluffs were too high and too steep. Nor do the streets that lie north of here go through; not until

FIG. 2.2B. DETAIL OF GLOVER'S *BIRD'S-EYE VIEW OF THE CITY OF SEATTLE.* The University of Washington is the largest building in the image. Arthur Denny's house sits alone on the corner of Front and Union, just above the cribbing wall. Yesler's wharf extends into the lower right side of the image. The panoramic photograph was taken from the end of that wharf.

Wall, about a half mile north of Union, does a street extend up from the waterfront to First Avenue.

In addition to restricting travel, or probably because of that, the bluff led to a key socioeconomic facet of Seattle society: the marked separation between the haves and have-nots. This dynamic shows clearly in two documents from 1878: a two-part panoramic photograph and a bird's-eye-view lithograph. The panorama covers the waterfront from Columbia to Pike Street, about a half mile in distance. The bird's-eye lithograph illustrates the entire city, with Lake Washington and the Cascade Range as background. (Bird's-eye views were a popular way to depict towns in the last half of the nineteenth century. Typically drawn by itinerant artists, the lithographs provided a highly accurate view of the urban infrastructure. Historian John W. Reps has written that, "better than any other single contemporary record," these images help us see how towns "grew and changed and how they appeared to residents and visitors alike.")[11]

Below the bluff was where the city's disenfranchised people dwelled; atop was where the more established citizens lived. For example, just east of and above where I stand, stood Arthur and Mary Denny's elegant house, at the southeast corner of Union and Front Streets. (Front is now known as First Avenue. I refer to it as First in discussions of the present and Front Street when it was Front.) Like many homes in the city, the Dennys' was two stories high and covered in white clapboard, and it had a large, covered front porch. Although a smattering of houses dotted the south and west slopes of Denny Hill to the north, the Denny house in 1878 marked the northern edge of the densest part of Seattle, which had tripled in population since 1870, to about three thousand people. Most homes were between Union and James from Front to about Fifth, though no area was particularly crowded. Many homes had large yards, some of which look a bit wild and many of which held fenced-in cows, goats, and chickens. In the modern landscape of skyscrapers that now covers this real estate, the few animals that live here are pigeons, perhaps a pet dog or cat, and an abundance of rats, certainly a far less edible menagerie. By 1878, people had begun to build east of Fifth, but they were outliers, like those in Belltown on the west slope of Denny Hill.

None of the homes in the main part of town were on flat terrain, and many were on hills far steeper than today, which meant that they would have had gorgeous views west to Elliott Bay and beyond. Getting to the buildings does not look easy, since the sidewalks were often wood, if they existed at all, and the steep roads were dirt and often not completely level. In rainy weather, the roads became morasses that made wheeled travel nearly impossible.

Scattered among the homes were institutional buildings. These include the ornate University of Washington (completed in 1861) and seven churches, including those of the Methodists, Presbyterians, Baptists, and Catholics. Sitting high on a knoll, the seventeen-year-old university building loomed large over the city. In contrast, most businesses were located near Yesler's mill and along the waterfront. To get to work, people walked or traveled by horse; the first street-trolley service didn't arrive until 1884.

Seattle in 1878 was clearly still a maritime city with its eyes and economy focused on the water, though the first glimmers of its future reliance on trains can be seen at the north and south ends of the bird's-eye view. To the north, a Seattle Coal and Transportation Company train emerges out of the woods south of Lake Union, bringing coal to a massive wharf extending into Elliott Bay at Pike Street. At the south end, two trains of the Seattle and Walla Walla Railroad and Transportation Company appear to be on a collision course as they head to sawmills and a dock built out onto the tideflats of the Duwamish River. These were the only two rail lines in the city.

Life below the bluff was less grand. "Persons living down there went through a good many hard times," said Agnes Lucas in a 1955 interview. She had lived in one of the shacks in the 1880s and remembered the landlord saying that he'd take away the windows and doors if her mother didn't pay rent. "He meant it, too." Agnes and her sister collected mussels on the beach and foraged in the woods for hazelnuts. To earn pennies, the girls, both under the age of ten, carried a beer bucket to the local brewery and got it filled for neighbors. Heat came from driftwood.[12]

In the panoramic photograph, a dirt path leads from Union Street down the bluff to several small homes clustered along the beach. To the north, more structures cling to the steep face. What we cannot see but can glean from other photographs, maps, and reminiscences is that these "beachcombers," as they came to be called, derisively, formed an extensive but tenuous community centered on the ravine at Bell Street.[13] Most lived in cabins or shacks typically made from whatever the builder could scavenge. A cabin owned by Old Billy Hoffner, a ship caulker who lived in it for at least a dozen years, had windows, canvas siding, and a roof made from old ship parts. Another resident used a stove made of bricks and sheet iron.

An article in the *P-I* about the beachcombers mentioned a dozen languages spoken by groups as diverse as Greeks, Italians, and Danes, though the reporter used more stereotypical descriptions. Many Native people continued to find this beach one of the few areas in the city where they were tolerated. Among the beachcombers, the men worked as fishermen, longshoremen, and laborers, while the women "stand at their doors and gossip." The children played on the beach and clamored eagerly for the peanuts tossed to them by the reporter. That the beachcombers were even noticed by the *P-I* man is somewhat unusual, since the papers and many of the citizens tended not to acknowledge such gatherings of

FIG. 2.3. HENRY AND LOUIS PETERSON'S PANORAMIC PHOTOGRAPH OF SEATTLE, 1878. Technically this is not a true panoramic image of Seattle. I refer to it that way because even though the two photographs do not align perfectly, they still provide an accurate panoramic image of Seattle. Woodward Grain House is the white three-story building sitting below Front Street. Denny Hill is the high mound with a handful of trees on it. Courtesy of Seattle Public Library, Image spl_shp5123 and Image spl_shp5125.

marginalized people, many of whom were forced by ordinance or circumstance to dwell on some of the city's least habitable lands.[14]

When they did notice the beachcombers, residents tended to complain about them. The shanties, lean-tos, and sheds of the former were a "blemish on this fair and growing city," wrote one observer in 1892. Others petitioned the city health officer to destroy what they called Shantytown, home to around five hundred people. Little came of their efforts. Instead, Shantytown residents had to worry more about landslides, one of which "caused considerable consternation" in this "most squalid" district. Because most of the residents were squatters, neither the city nor the landowners would do anything to prevent further landslides.[15] In

contrast to those forced to live on the tideflats and beaches, businesses that moved their operations onto these territories were often portrayed as entrepreneurs. For example, those who built a long wharf jutting out into Elliott Bay were called "enterprising citizens."[16]

◆ ◆ ◆

For the next few blocks south of Union Street, the historic shoreline runs down Post Alley, between First Avenue and Western Street. The alley's name came from a post office that opened in June 1880 at the corner of what was then Mill Street, now Yesler Way. (As I do with Front and First, I refer to Mill Street when it was Mill Street and call it Yesler in discussions of circumstances after 1895, when its name was changed.) Long associated with the Pike Place Market, Post Alley—which has achieved notoriety for Seattle's most disgusting tourist attraction, the Gum Wall—passes by two spots that provide graphic insights into the topography that concerned the city's earliest settlers.

I am pretty sure that the first location was not planned as a clue to our past: the Harbor Steps, a mixed-use development that combines shops, apartments, and restaurants with a parklike space of concrete and waterfalls. Although there is no evidence of waterfalls having existed anywhere in the city, seeps and springs were widespread in early Seattle and are a sure sign of the contact between the Lawton Clay and the Esperance Sand. For example, there were at least nine springs providing drinking water for the early residents clustered in the downtown urban core. The most famous one we still honor in the name Spring Street. Concrete and buildings have covered most of the city's seeps and springs, but the water persists in unexpected places. Go to any park with steep bluffs, such as Frink, Golden Gardens, Schmitz, or Ravenna, and you will find running water pouring over ledges, cutting into hillsides, and providing habitat for an unusual array of plants, such as devil's club, horsetail, and nettle. The waterfalls of Harbor Steps may be completely artificial, but you can assume that when Denny, Boren, and Bell canoed across Elliott Bay, seeps and springs would have been commonplace along the bluffs and slopes that rose up from the waterfront.

Farther south, at Seneca Street, is the second spot that offers graphic insight into the topography. This is the southernmost point at which an east-west street does not go through from First Avenue to the waterfront. From here south, the streets do not, and did not, have to contend with a high bluff along Elliott Bay. It may not be obvious, but if you follow First Avenue from Lenora you drop about eighty-five feet. Those original high bluffs that started in the north steadily became lower, though they still created a barrier and led to Seattle's first official, city-government-sanctioned reshaping of the topography.

◆ ◆ ◆

To understand how this barrier influenced early development in Seattle, I turn again to the 1878 panoramic photograph. Henry and Louis Peterson took the pair of shots looking east from the end of Yesler's Wharf, which extended from the original mill site at what is now Post Avenue and Yesler Way into Elliott Bay and turned north. Dominating the middle of the panorama is a three-story building with wood siding, known as the Woodward Grain House, home to what sounds like a trifecta of fun and profit—Peter's Furs, Cigars and Liquors.[17] The Woodward sits on the west side of Front Street, between Madison and Marion, at what is now the location of the older of the two downtown federal buildings.

Directly behind the Woodward rises a wall of stacked horizontal logs, the top of which is equal in height to the top of the Woodward's second-story windows. The ends of abundant, randomly placed logs stick out of the wall by several inches to several feet. These logs, up to thirty feet long, have been driven into the hillside and help form a secure framework designed to hold the hillside in place.

Such a structure is known as cribbing, and it was, and still is, widely used to hold back unsafe slopes. This cribbing was Seattle's first attempt at blockading the tides of Elliott Bay, at least from Madison to Columbia, which is where the timber wall stands next to the water. Built during an 1876 regrading project to make Front more accessible, the wall continued from Madison to University, becoming more of a retaining wall holding up Front than a seawall holding back Elliott Bay.

The regrading of Front Street was a critical point in Seattle history. Not only was it the first official effort at rebalancing the terrain in favor of the citizenry, but it also exemplifies the many relatively small-scale street regrading projects to follow, which can be felt throughout Seattle. Look at any street in the city and you can assume that some sort of engineering took place, meaning that before the road's existence, the ground surface probably undulated more, and was steeper, than the surface of the eventual road. This is particularly true in the downtown core, where most of the original steep and undulating hills were tamed and then covered with much-easier-to-travel roads. Regrading, like that of Front Street, has played a critical role in the development of any urban landscape.

The Front Street project began on June 8, 1876, when the city council passed Ordinance No. 112, *To Provide for Grading Front Street*, which described the bidding process, specifics on what would be done, and who would have to pay for the work. As happened with all later regrades, the cost—from $5.00 to $340.19 per lot—was paid by landowners whose property bordered the street. Front Street was chosen first because of its location along the shoreline. A more level roadway with slighter grades would make it easier for businesses to move north from the center of commerce at Mill and Front Streets. In addition, the new cribbing would enable better access to docks and to buildings such as the Woodward.

When the council opened bids in July, George Edwards won the right to cut and fill Front Street from James Street to Pike Street with a bid of $9,000.[18] After getting his contract from the city council, Edwards began grading on July 13, 1876, in front of Henry Yesler's property between James Street and Cherry Street, and just north of Seattle's commercial core. The city's work plan required Edwards to reduce grades as steep as 13 percent to under 7, and typically less, since Front Street rose in a little over a half mile from 12 feet above sea level at James to 107 feet at Pike. To keep back Elliott Bay, Edwards would have to build the permanent log cribbing—up to 27 feet high—along the western edge of Front. The project would require twenty-six thousand cubic yards of dirt, with excavations along Front providing two-thirds of the material and the rest coming from cutting down the side streets running east from Front.[19]

Edwards employed up to ninety men and twenty mules, oxen, and horses. Using picks, shovels, carts, and wagons, the workers made their deepest cut, of twenty-five feet, at Spring Street and their greatest fill, an addition of eight feet, between Marion and Madison. The extensive cribbing built along the shorefront was composed of cedars "hauled from the woods back of Belltown," then more or less a suburb of Seattle.[20] Edwards's crews raised many houses on stilts, which would eventually be supported by fill, and left others on the hillside high above the now lowered road. This led to homeowners, such as Arthur Denny, having to build what was described as a "prevent wall," to prevent their houses from sliding.[21]

Work did not always go as planned. Edwards had to rebuild an extended section of cribbing because he did not use logs long enough to anchor the cribbing into the embankment. His initial logs were just ten feet long; later ones were thirty feet. When the rains came in October, runoff washed away fill, cut deep gullies, and caused sliding. In some cases, the streets were "literal cesspools," reported the *P-I*.[22] By January, work costs had increased to thirteen thousand dollars. Seattle's typical winter rains continued to hamper progress, causing delays that extended into April.

Finally, on May 24, 1877, the *P-I* could report that the work was completed. Front Street was now "one of the most pleasant and excellent thoroughfares to be found in any city this part of the coast."[23] To the editors of the *P-I* and Seattle's other newspaper, the *Daily Pacific Tribune*, grading was essential to what would soon be labeled the Seattle Spirit. An editorial in the latter noted that "any one with half an eye can see the good already accomplished. . . . [It is] stamping the growth and business of the city as the most enduring, desirable character."[24] To the *P-I* the grading "bespeaks enterprise; shows that we mean business; that we mean to stay here."[25] Just twenty-five years after its founding, Seattle was a city to reckon with.

And once local leaders realized that they could subjugate the topography, they regraded with zeal. On Mill Street, workers made cuts of up to twenty feet where the road climbed east. One block south on Washington Street, the men dug up stumps and burned huge logs in order to push that ever-extending road farther east. Within a few years, the city council passed ordinances for grading Second, Third, Fourth, Madison, Marion, and Commercial Streets.

◆ ◆ ◆

Back in the present, I walk along First Avenue to Columbia, turn right, and head down to Western Avenue, to one of my favorite areas illustrating how Seattle's early landscape still influences our modern city. This section of Western Avenue looks to be an engineer's nightmare. The middle of the street is higher than the edges and the entire road surface undulates. The concrete curbs look as if they had been poured by a drunkard, sometimes disappearing under the street and sometimes rising ten inches above it. Near the southern end of the street is a low point that has held a pool of water every time I have walked by it.

Worse than the uneven pavement, though, is the condition of the building at 619 Western Avenue, on the west side of the street, where a jagged vertical crack runs from the rooftop down behind the fire escape for nearly three stories. Additional cracks, some up to eight inches wide, also split the east and south exterior walls, as well as many interior walls. The building at 619 opened in 1910, one of a pair of adjacent warehouses (the other is the Polson Building) built to take advantage of Seattle's expanding shoreline. Trains and delivery vehicles accessed the warehouses from Railroad Avenue, which was slowly extending the shoreline into Elliott Bay. A photograph from 1917 featuring painted signs on the building illustrates how the economy has changed—signs advertise sailmakers, wagons, and ship chandlers (or supplies)—and how it hasn't: others advertise coffee roasters, pancake flour, and canned fruits.

The cracks and the irregular street surface have resulted from what lies beneath. In 1996, the Washington State Department of Transportation drilled a core nearby as part of a seismic-vulnerability study of the Alaskan Way Viaduct. In the first four feet of the core, the drillers found three inches of asphalt; three inches of railroad ballast, or gravel; twelve inches of concrete; and eighteen inches of broken concrete, pieces of creosote timber, and rounded gravel. The technical report notes that the drillers then reached organic soil and "a void from 4.0 ft. to 8.0 ft." before hitting moist, loose soil again. Fourteen and a half feet down, the core changed to pieces of creosote timber and a new feature, an aroma of rotten egg and sulfur. The remaining eight and a half feet of core is described as contaminated organic soil. A final note adds, "See samples at own risk."[26]

When I asked Bob Kimmerling, the engineer who prepared the technical report, about these notes, he reminded me that this was near the location of Yesler's original mill. "They dumped a lot of sawdust and other material, so it's not surprising that it's rotting down there," he said.[27] Nor is it surprising that the road surface, and many sidewalks in Pioneer Square, are less than stable with such a stew of random ingredients making up the Swiss-cheese texture underground. Another engineer told me that the void is probably a buried structural element, such as a decommissioned utility vault, a buried sidewalk, or a foundation element that was part of a structure. There is probably some sort of wall or piling supporting the "roof" of the buried object, but it was simpler to just build around these structural elements than remove them; and once buried, they would be covered with more fill.

When Seattle's pioneers started to expand the city and transform its challenging topography of seaside bluffs, steep hills, and unlevel land, one of their basic practices was to drive a double row of pilings out from the shoreline, lay timbers across the tops of the pilings to form piers and wharves, and build out atop the wood. They could then dump material under these structures, undertaking the land-making process known as wharfing out. And this is what the cores show: cinders from coal bunkers and steamship loading areas; ceramics, glass, and metal from homes and businesses; pilings and planking from street supports and wharves; charcoal and ash from the rubble of the Great Seattle Fire of 1889; bricks that formerly surfaced streets; and sawdust and wood from mills, which generated a horrible aroma as they decayed.

Little of the subterranean detritus reveals information beyond what was previously known, but taken together, the cores reexpose Seattle's early history of industry and commerce, which the advancing waves of progress covered: coal from Newcastle and points east, wood from the ever-shrinking forests that surrounded Seattle, brick from clay beds along the Duwamish, transportation by sea and later by rail.[28] All buried underfoot and out of sight.

The cores vividly illustrate how much Seattle's early citizens relied on the area's natural resources. Like many others around the globe, the pioneers saw those resources as something to use for their own needs. Why shouldn't they have? They couldn't imagine that anyone could dig out all the coal or exhaust the forests. Those pioneers were dreamers and planners and doers. They wanted to make Seattle a great city, and nature wasn't going to stop them. Either use it, sluice it, drain it, drill it, or fill it. Whatever was necessary.

The building at 619 Western also suffers from another buried topographic story of Seattle. Tidal water regularly creeps under the building and into its foundation, despite its location more than 250 feet from the city's seawall. The seawall does do its primary job of keeping Elliott Bay at bay; but as the tide rises and falls, water percolates through the porous wall. Major repair work on the building led to the discovery that this inundation causes the level of the groundwater to vary about five feet, leading to regular submerging and exposure of the underground wood pilings that support the building.

In a well-built building, a concrete casing caps the end of each wood support pile and the building rests on the concrete. Builders originally followed this practice when they erected 619 Western, but subsequent exposure during low tide rotted out the wood so that the concrete caps no longer contacted the pilings. Combined with its structurally compromised underground garbage heap of support, the rotted pilings resulted in 619 Western's network of gaps and fissures.

You can no longer see these cracks. In 2010, the Department of Transportation proposed to destroy the building because engineers didn't think it could withstand the vibrations of the new State Route 99 tunnel going under the building. A last-minute fight by preservationists led to the department reversing its plans, and in 2011 it decided to rehabilitate the building. Construction crews placed more than 250 new concrete piles, in addition to giving it a complete seismic retrofit, which can be seen in the massive steel beams forming *X*'s behind the windows on the front of the building.[29]

The new, modern, concrete-and-steel seawall, which is slated to be finished in 2016, will allow water to percolate through, like the old wall did. As a spokesperson for the project told me, "You cannot stop Mother Nature." High tides and storm surges will always be a factor and water will always need to infiltrate the seawall. Designers didn't want to create a barrier for both freshwater that flows down from the slopes above and from salt water that has made it through the wall. As happens with the old seawall, the new one will allow groundwater and seawater to weep out of the wall and into Elliott Bay.

In many ways, 619 Western is the new face of how modern Seattle addresses its landscape. Those who put up the building had little idea of the consequences of building on made land. They knew to put in pilings but had no way of understanding the local geology, or even the effects of tides and groundwater. Armed with more thorough knowledge of the geology of our location, we now take that information into account when we decide to retain and maintain older structures with new construction, and when we undertake major projects such as the new seawall and the State Route 99 tunnel.

◆ ◆ ◆

Continuing my journey, I turn left at Yesler Way and walk over to First Avenue and one of the most important geographic locations in the city. To the Duwamish people, this was and is *Sdᶻ idᶻ əlʔalič*, or Little Crossing-Over Place, a reference to a portage connecting a lagoon to a trail to Lake Washington. As many as eight large longhouses stood here and held around two hundred people, though only one structure remained when the Denny party landed in 1851.[30] This did not mean that the Native residents no longer lived here, however; many had died from disease, and many had moved to other locations around the area. More than simply a portage, this small site defined the land for the Native people; up to at least the 1940s, tribal elders across Puget Sound used the name Little Crossing-Over Place instead of the more modern designation of Seattle.[31]

To the new settlers, this area was the heart of the business district, starting with Henry Yesler's mill in 1852 and continuing for more than half a century. The importance of this intersection can be seen in the buildings around it. Bearing the names of early residents and entrepreneurs, such as Denny, Terry, Yesler, and Schwabacher, most of the elegant, multistory structures of sandstone and brick were built within a year or two of the Great Fire of 1889. They were Seattle's sign that it had united under the banner of Seattle Spirit and risen from its ashes. But this intersection also divided the city.

Yesler Way marked another informal line between Seattle's more established community and its second-class citizens, or the people who became known as those who lived on Skid Row.[32] South of Yesler Way were the city's cheap hotels, boardinghouses, and tenements, occupied primarily by single, transient men, many of whom worked in the often-dangerous natural resource industries that drove Seattle's economy. Serving these men were a host of saloons, brothels, and "box houses," low-class theaters where the actresses provided more personal, postproduction entertainment.

First and Yesler is also one of the most curious geographic locations in Seattle, the spot labeled as the "Neck" on Phelps's 1856 Seattle map. It was about 200 feet wide by 250 feet long and is what is known as a tombolo, or a sand or gravel bar that connects a mainland to an island. There is some speculation about this feature. Was it always above water or did it periodically disappear under high tides? The earliest map to give a hint of the mutability of the tombolo was one drawn by a member of the United States Exploring Expedition of 1841, under the command of Charles Wilkes. The best-known map from the Wilkes exploration of Puget Sound illustrates the region from about Port Townsend down to the southern end, or what we now call Olympia. What would become Seattle is merely an unnamed headland projecting into Elliott Bay, which Wilkes most likely named for Midshipman Samuel Elliott but possibly for the expedition's chaplain, Jared L. Elliott.

The expedition atlas includes a second, close-up map of Elliott Bay. This more detailed chart depicts Elliott Bay ringed by conifer-covered bluffs bisected by gullies. Seattle is labeled as Piners Point,[33] though it looks more like an island or high mound covered in shrubbery. North and northeast of the mound, the cartographer has drawn in what looks like a marsh with tufts of plants. The marsh gives way to tidal flats east of the mound. There is no sandspit as shown on the Phelps map. Nor is there a tombolo.

Were these features underwater when the Wilkes crew made the maps? Did they overlook them, or had the features not yet formed in 1841? We can never know, but the available evidence does lend credence to the idea that the area known as the Sag, or the Sawdust, was low enough that high tides created an island out of the high mound just south

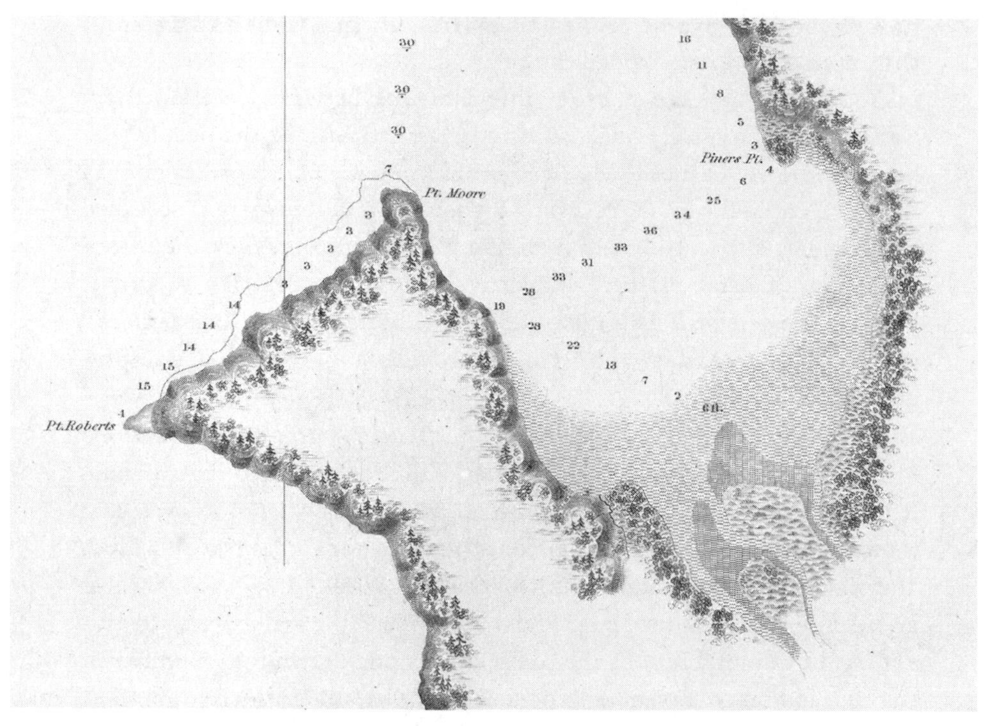

FIG. 2.4. DETAIL OF CHARLES WILKES'S ELLIOTT BAY MAP, 1841. Detail of Elliott Bay, from a map created by the United States Exploring Expedition, 1841. This was the first map to provide an accurate rendition of the mouth of the Duwamish River and the future site of Seattle—or Piners Point, as it is labeled here. The map implies that the point, which appears to be a shrub-covered mound, could become an island when high tide covered the marsh surrounding it. Note how close to shore Elliott Bay becomes deep water, well suited for large vessels. Library of Congress, Geography and Map Division.

of the Neck. Two cores drilled in 2002 at the location of the tombolo back up this assessment. Both show thick beds of loose gray sand, up to ten feet thick, exactly what one would expect to see in a tombolo. These sand layers are six to thirteen feet below the present-day road surface.

Although the low part around Yesler's mill was filled quickly, the eastern inlet, which fed water into the marsh behind the island, took more than thirty years to shut. The mound did not have an official name; some called it Maynard Town or Maynard Point, after Doc Maynard, who settled there, or simply the Point. Many recent histories of Seattle call it Denny's Island, although Arthur Denny neither owned the land nor lived there.[34] It covered about eight acres, or was about half again as big as Freeway Park.[35]

Standing at this intersection, contemplating the unstable nature of this spot, I have to wonder what Denny and his fellow settlers were thinking. Were they so focused on having a good port that they failed to fully evaluate the land around them? Was it desperation that prompted their decision? They had traveled more than halfway across the country to establish a new town, and they didn't have many choices; this was one of the few spots in Elliott Bay with some level land that didn't sit at the base of a steep bluff. Or were they influenced by the Native longhouses, figuring but never apparently acknowledging that those who had lived here for centuries knew the best places to build? Although the answers are lost to history, Denny, Boren, and Bell did make a good decision, probably through a mix of luck, drive, and timing.

◆ ◆ ◆

One of the hardest geographical aspects of Seattle to appreciate is the fact that once you are south of Yesler Way, nearly everywhere you walk you are traveling well above the city's original street surfaces and, in some places, atop the island mound. If you want to get a feel for this topographic quirk, visit the southwest corner of the Pioneer Building, the grand building just east of the pergola in Pioneer Square. Look at the granite wall above the steps at what was a former entrance. There you will find the carved words and numbers "City Datum, Elev. 18.79."

The City Datum is the zero-elevation-point reference for the city, or the point on which all other elevations are based. Another way to consider this—and it did happen in the early days of Seattle—is that without a standard, adjacent property owners might end up with corners several feet apart, both horizontally and vertically, because they didn't start from the same elevation base. In Seattle, zero is sea level; so when you stand at the steps of the Pioneer Building by a two-inch-wide square chiseled into the granite, you are 18.79 feet above what was determined to be sea level when the city set this datum in stone in 1891.

Engineers in 1875 determined the first city datum by setting up a tide gauge on one of the downtown wharves, measuring tides over the month of June, and using the mean of the high tides as zero.[36] After determining this level, the engineers had to transfer the information to a permanent location where later surveyors could find it easily. Accordingly, they chose one of Seattle's first granite buildings, the one-story-tall Dexter Horton Bank on the corner of Washington Street and Commercial Street and established Seattle's initial datum at 8.835 feet below the level of the top of the lower granite step in the doorway of the bank.[37] After Mr. Horton's bank succumbed to the Great Fire of 1889, when twenty-nine blocks of downtown burned to the ground, the datum point was moved to its present location.

If you return to First Avenue, turn left, and walk south one block, you can see where Dexter Horton's bank once stood. It is now the sandstone-and-brick Maynard Building (completed in 1892), with its off-center arched entry. Clearly you did not lose eleven feet in elevation while walking one block. Instead, the difference between the two datum points reflects the variation between the present road surface and the historic road surface. In fact, you are standing eleven feet higher than someone standing at the same intersection in 1875, which is why the Maynard has what looks like a formal entrance and steps that lead down to the building's basement-level businesses.

When Seattle's citizens started putting up homes and businesses, they built directly atop the land surface. In the area of the Neck and the Sawdust, where the Dexter Horton bank stood, this meant that they built just above sea level, on the sawdust generated by Henry Yesler's mill. (The Pioneer Building is on the site of Yesler's home.) This practice continued until the Great Fire, when city officials decided to change Seattle from what the *Seattle Times* described as "misshapen and ugly" to "shapely and beautiful."[38]

At the time of the fire, Seattle's population had exploded, reaching about thirty thousand.[39] The rapidly growing city now had its first five-story building, eight sawmills, and—to ensure that everyone was happy—four breweries and five candy factories. Commerce still centered on the intersections of First, James, and Yesler but had also spread to the shores of Lake Union and out to the growing towns of Ballard and Fremont.

Arthur and Mary Denny's house would, by that time, no longer appear to be the northern edge of the city as it had in 1878. By 1889, homes covered Denny Hill and had begun to climb up Queen Anne and First Hills, which were quickly becoming the most fashionable places to live. There were also houses along Lake Union, atop Capitol Hill (though the latter had yet to be named), and overlooking Lake Washington. To reach these homes, residents could ride one of several street trolleys that spread across the city. As one historian writes when comparing 1880 to 1889, "Instead of a town, we would see a city."[40]

Much of this progress succumbed to the fire, but to many in the young city this meant new opportunity. Central to Seattle's new look would be the raising and widening of the streets in the business district that lay in what is now the Pioneer Square area. An editorial in the *P-I* observed that the "narrowness and crookedness of the streets" had regularly led to blockades of wagons and carriages before the fire. With a clean slate, planners could alleviate this early-day equivalent of rush-hour traffic.[41] The city council called for the expansion of Second and Third Streets by taking twelve feet from the lots on each side, which would make these thoroughfares ninety feet wide. Front and Commercial would grow from sixty-six feet to eighty-four feet wide.

Just as critical to enlarging the channels for business to flow was establishing a consistent grade for the new streets, which translated to raising them. Most of the Pioneer Square streets would be raised from their prefire elevation by an average of five feet per block. Elevating the streets would address another flowage issue of utmost interest to citizens: it would improve the sewer system, which spewed directly into Elliott Bay, and promote surface drainage. New pipes would be embedded at a slight angle in the fill of the higher roads, providing a gradient for liquids and solids to flow downhill.

Work on the new roads was not easy. The city controlled the space from curb to curb, and the street-construction project required building retaining walls, finding material to fill in between the walls, transporting the material to the site, filling in the street, and finally paving over, usually with wood, the surface of the new, higher street.[42] At the same time, property owners began to build the necessary retaining walls at the curbs, because city regulations stated that individuals and not the city owned the space from the building to the curb.

Not willing to wait for the city to finish raising the roads, business owners also started to construct buildings, wanting to be open for business and to sell items to the bustling city, which had added over ten thousand people in the year or so following the Great Fire. Using stone and brick—a new ordinance-mandated requirement—builders located their first floors at the level of the old streets. Because property owners knew that the streets in front of their properties would soon be higher than at present, they used less expensive materials and paid less attention to the details of their temporary storefronts. The original sidewalks were also at the prefire level, which made it easy to enter the new business establishments.

Permanent sidewalks didn't appear until 1893, when the city council passed an ordinance prescribing the requirements for the new sidewalks that would be constructed at the level of the new streets. You can still find some of the old sidewalks in place; they are the ones with vault lights, also called sidewalk prisms, embedded in them. Flat on top and prismatic below, the glass lets light into the subterranean canyons, which were often, and still are, used for storage.[43] The prisms were originally clear and have since turned purple owing to high manganese content, which changes color during long-term exposure to ultraviolet light. There are still around nineteen thousand such lights dotting the sidewalks of Pioneer Square.

◆ ◆ ◆

Wanting to complete my plan to follow Seattle's historic shoreline, and to trace the edges of Maynard Point, I leave the Maynard Building at First and Washington and head to the alley that runs behind it and continues south. According to Phelps's map, I am still in the area of the Neck. I feel like I am on an unconsolidated surface, because the alley rises as I cross the sidewalk south and descends as I drop into its midpoint. The asphalt reminds me of the surface of baked brownies—uneven with bulges, cracks, and pockets—though a bit more pungent with its aromas of trash and urine. The redbrick building to my left looks as if the entire back end had settled slightly and had relieved the stress by bulging out at the base. I count fourteen tie-rods sticking out of the two-and-a-half-story wall, further signs that the structure sits on a settling subsurface.

If I had walked the same route in the early days of Seattle, I would have been ascending the mound south of the Neck, past a handful of homes and businesses, to reach a high point overlooking a sandy bluff that dropped down to a beach. This being modern Seattle, however, I walk by orderly rows of green compost and blue recycling bins. At what would have been the high point—the present-day Jackson Street and the alley—I would have reached a disreputable structure, the infamous Madam Damnable's hotel and brothel illustrated on Phelps's map.

Although Phelps's map does not indicate a bluff, we do know that Madam Damnable's sat perched above Elliott Bay. John Y. Taylor, assistant surgeon on the *Decatur*, sketched the Seattle waterfront and included a blockhouse and Madam Damnable's atop substantial bluffs rising abruptly from the water. Taylor's drawing also includes what look like steps, or a ramp, leading from the water up to the building. Judging from the height of the two-story building, the bluff looks to be twenty-five to thirty-five feet high.

Other evidence, however, suggests that the bluff was not that tall. During the preliminary studies for removal of the Alaskan Way Viaduct and for the seawall replacement, geoarchaeologist Brandy Rinck used a tool known as a geoprobe, which takes a core sample by pushing a pipe into soft sediment, to analyze the subsurface along First Avenue South. "I am guessing that the bluff on the west side was about ten feet high, but we cannot know how much till was removed by builders, so it may have been higher," says Rinck.[44] She adds that the west side of Maynard Point was probably steep because of the way the waves would have eroded it, unlike the more protected south and east sides, which likely sloped gently into the water.

Because of the modern street layout, I cannot easily trace the south and east sides of Maynard Point, so I end up heading directly over to a spot that I know was the southwestern edge of the salt marsh depicted on Phelps's map. I am on Second Avenue, between South Jackson and South Main Streets, at the vehicle entrance to the headquarters of the Seattle Fire Department. Directly east of me would have been the sandspit that extended west from the former shoreline below the coastal bluffs. At present, I have the pleasure of seeing a parking lot.

In his drawing of the marsh, Phelps has depicted four lagoons. All are amoeba-like, with no defined shape, though one of the two larger ones has two arms that extend northwest. Surrounding them he has drawn parallel lines sprinkled with four or five closely spaced vertical dashes meant to depict tufts of grass.

"I don't believe that particular pattern drawn by Phelps," says Hugh Shipman, a coastal geologist for the Washington State Department of Ecology. "I think he didn't want to get his feet wet, and what he drew was a stylized marsh."[45] There is also the possibility that since Phelps

FIG. 2.5. JOHN Y. TAYLOR'S SEATTLE PROFILE, 1855. The three-masted *Decatur* lies at anchor in the foreground of Elliott Bay. To the left of it, a brig waits at Henry Yesler's wharf, and behind the wharf, smoke billows from Yesler's sawmill. Note the high bluffs that run along the shoreline. The building at the far right is Madam Damnable's, where a ramp, or steps, leads up from the bay. What looks like a cabin is the blockhouse, or fort, built on a high point at Cherry Street before the Battle of Seattle. Courtesy Yale Collection of Western Americana, Beinecke Rare Book and Manuscript Library, Yale University, image 1079559.

was focused primarily on the military aspects of the landscape, he did not study the marsh and simply depicted it more artistically than realistically. Shipman tells me that nature tends not to produce multiple lagoons in such an environment. More typical is a single body of open water, rarely more than a few feet deep, which can drain out completely during low tide, leaving barren areas.

The exact border of that early-day salt marsh is pretty sketchy, mostly because not enough core samples have been taken to precisely define the edges. Instead, we have to rely on Phelps's map; though not perfect, it provides a reasonable outline, as shown on the accompanying map. In addition, unlike Maynard Point, the eight-acre marsh had daily and seasonal rhythms that resulted in constant change.

Where the marsh does show up in core samples is in the few taken from around Occidental Park. In one there is a four-foot-thick layer of sawdust and wood chips. At another the sawdust is only two feet thick but is buried under twenty-three feet of clay, silt, sand, brick, mortar, asphalt, cinders, concrete, and lumber. A third contains pieces of wood redolent of creosote and petroleum that form a poorly consolidated layer more than ten feet thick. As has happened with so many wetlands, this one became a garbage pit, a place to bury trash and history.

Other clues to our vestigial landscape are more graphic. Throughout Pioneer Square and its surroundings, sidewalks sag, alleys buckle, foundations sink, and walls slump. These are the signs visible to the public. Ask building owners and tenants, and they will tell you of another problem—basement flooding. It is surprisingly common, particularly during winter, when high tides can force water into buildings. One building owner told me that high pressure would force water up through hairline fractures in his foundation, where it would shoot out like mini fountains. The only way he keeps his building basement dry is with sump pumps.

We may have buried Seattle's earliest topography, but we can never escape its influences. It is sitting just below the surface, regularly popping up, reminding us of our origins and, perhaps, keeping us a bit humble.

As I finish my exploration of Seattle's historic downtown shoreline, I cannot help but notice how unnatural Pioneer Square feels. The entire neighborhood is pancake flat, a sure sign in glacially carved Seattle that something has changed. Few other unaltered spots in the city, and few altered ones, for that matter, look or feel like the made lands around the former marsh and mound; for no matter where you stand, you will soon encounter topographic change. You will have to ascend a hill, cross a gully, descend a ridge. Not around Pioneer Square; it is a landscape ironed flat, with neither wrinkle nor crease.

In the postfire city, consumers wouldn't have to worry about straining themselves on hills or read tide tables to know when to leave home or use the bathroom. Horse-drawn deliveries could move about effortlessly, tromping along level alleys and streets. Railroads could easily access the wharves starting to dot the waterfront. The wheels of commerce could roll on in, find their way around, and never be slowed by topography.

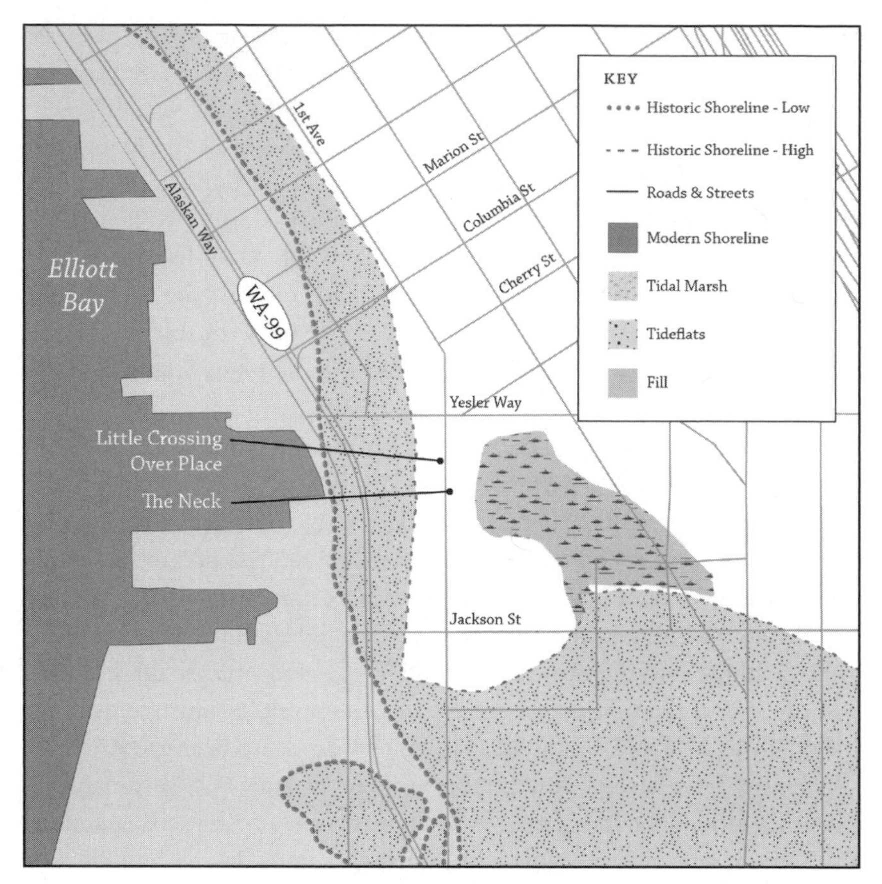

MAP 2.2. LAGOON. Historical mapping by UW's Puget Sound River History Project, adapted from the Waterlines Project.

Like so many successful pioneers before and after them, Seattle's thriving business owners wanted to show that they had moved beyond their humble beginnings. No longer was Seattle an isolated village of a few dozen, built on wheelbarrows of sawdust. The Seattle of 1890 was thriving. In just ten years, its population had exploded from 3,533 to 42,837. In another ten, the city would double again. Its citizens clearly viewed Seattle as the premier city of the state.

Even more drastic change was about to come to Seattle: the railroad. The first train, covering a little more than a mile, between Lake Union and Elliott Bay, had started running in 1872; and by the time of the Great Fire, rail linked Seattle to the rest of the country. But the service was spotty at best; passengers detrained at shacklike depots—that is, when the trains successfully navigated the skeletal trestles that crossed the tideflats south of downtown. Within three decades of the fire, however, Seattle had two beautiful passenger train depots with daily transcontinental services, several freight yards, and a train tunnel cutting under the city. All these changes eventually led to an even greater alteration of the landscape that would permanently alter the core of Seattle and produce the topography of the modern city.

CITY DOCK

3

Filling in the Duwamish River Tideflats

AS AN URBAN EXPLORER, I DON'T ALWAYS END UP IN THE CHAMber of commerce–approved destination spots of Seattle. Right now, I am engulfed in the jarring waves of noise emanating from Interstate 5, which is about 250 feet east of where I am standing amid aluminum siding and cinder-block warehouses. Adding to the pleasure are long stretches of chain-link fence topped by barbed wire, which protect such valuables as wooden pallets, fifty-gallon drums, steel dumpsters, and in one case, the most extensive array of forklifts—in various states of readiness and repair—I have ever seen.

I have come to this side lot of Seattle's old industrial heart, far away from the modern hipster city, because of what runs up the alley between the many graffiti-tagged businesses: parallel steel rails of a train track. No trains run on this route; and judging by the blackberries growing between the rails, none have in a long time. But this is a historic spot and another link to Seattle's early attempts to reshape its topography, for the long-abandoned tracks run along what is the oldest extant rail right-of-way in Seattle. It dates back to May 1, 1874, what one newspaper described at the time as "the most remarkable day in the annals of town."[1]

On that lowery day, all of Seattle gathered near where I now stand, between Ninth Avenue South and Airport Way and a bit north of South Dakota Street.[2] The citizens of the young town had come to build a railroad that would be

View of the Duwamish River tideflats from Beacon Hill, 1884 (detail)

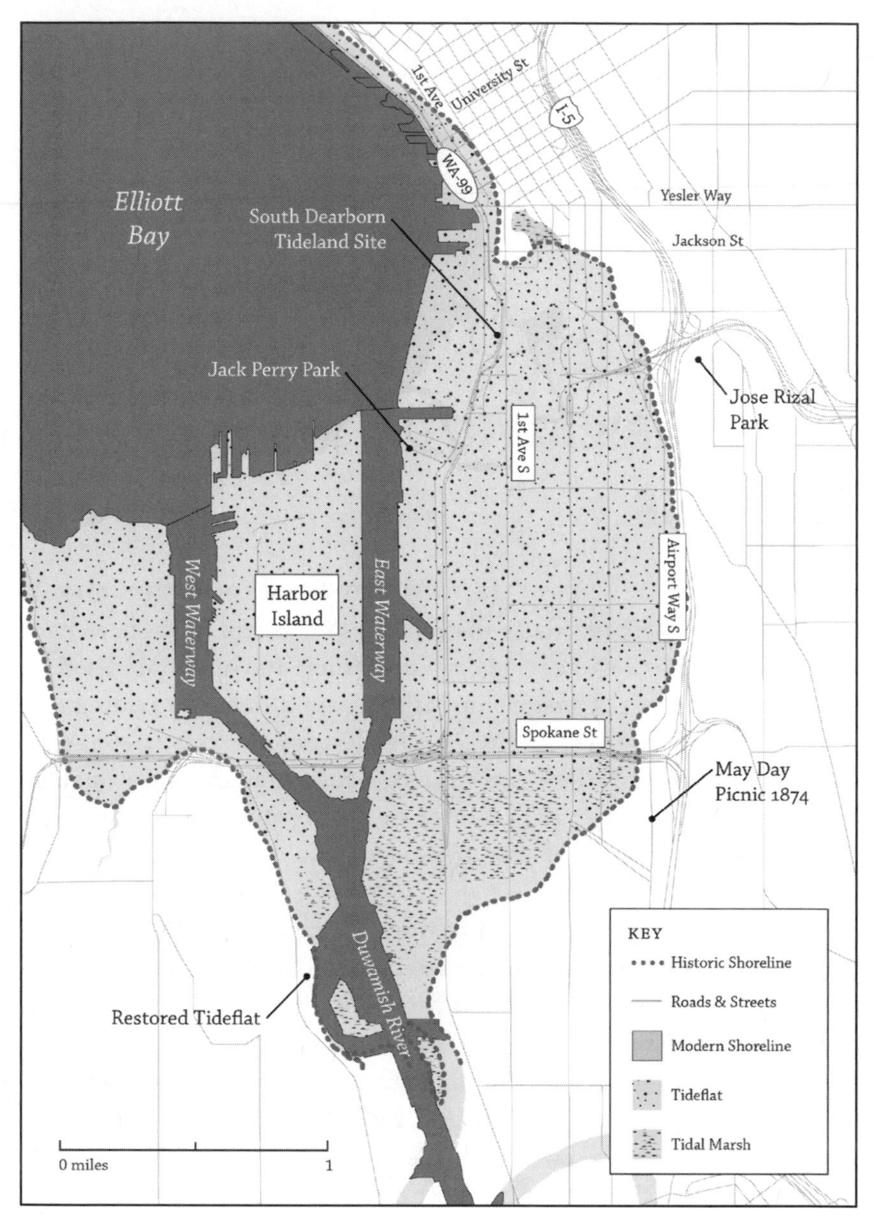

MAP 3.1. DUWAMISH RIVER TIDEFLATS. Historical mapping by UW's Puget Sound River History Project.

their ticket to the future. It would be the first step in making one of the greatest changes to Seattle topography, the conversion of more than twenty-two hundred acres at the mouth of the Duwamish River from tideflats to made land.

Almost a year earlier, on July 14, 1873, Arthur Denny had received a devastating one-sentence telegram: "We have located Terminus on Commencement Bay."[3] It had been written by the directors of the Northern Pacific Railway. Not only would the great train company not come to Seattle and finally connect the city with the rest of the country, but it also would end its cross-country track in Tacoma, Seattle's most-despised neighbor.

Over the previous decade, the young and scrappy towns on Puget Sound had fought for the privilege of being the final destination of the Northern Pacific's transcontinental route. Each town had wined, dined, and kowtowed to surveyors and executives, hoping that the railway men would recognize the virtues of its location as a site for the Northern Pacific's terminus, forever putting their town on the map. Seattle's ever-hopeful citizens had pledged three thousand acres of land, forty-eight hundred feet of waterfront property, 750 town lots, fifty thousand dollars in gold coin, and two hundred thousand dollars in bonds, at a time when merely eleven hundred people lived in Seattle; Tacoma could claim only two hundred residents.[4]

Despite offering their soul, their land, and their money, Seattleites could not purchase the Northern Pacific's favor. Since the railroad owned much of the land around Tacoma, it stood to gain far more when the town grew into its terminus status. Tacoma's triumph was a blessing in disguise for Seattle. No longer would Seattleites have to be the "serfs of a soulless monopoly under the direction of a hostile power," editorialized the *Puget Sound Dispatch*, though many of Seattle's old guard would remain enthralled by distant wealth. Just as important, the paper said, the city would retain better control of its own destiny.

It made use of that control almost immediately, on July 17, 1873, in the "largest and most earnest meeting of citizens ever assembled."[5] To hell with the Northern Pacific, the crowd snorted with petulance and pride: the town would build its own railroad. A week later, newly elected commissioners filed articles of incorporation for the Seattle and Walla Walla Railroad and Transportation Company. The end point of the tracks would be Walla Walla, 289 miles away through foothills, over mountains,

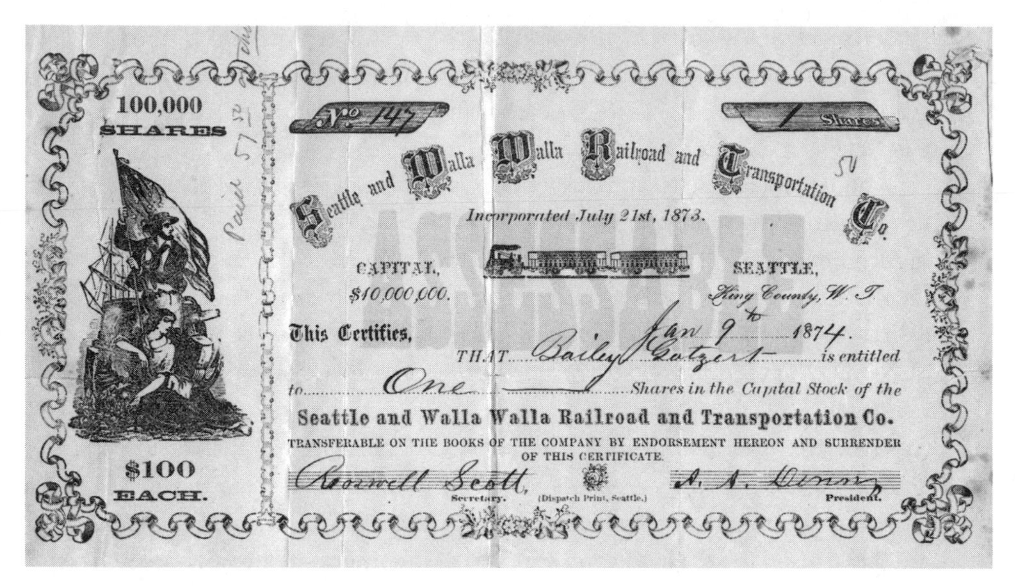

100,000 SHARES

No. 147

1 Shares

Seattle and Walla Walla Railroad and Transportation Co.

Incorporated July 21st, 1873.

CAPITAL, $10,000,000.

SEATTLE, King County, W. T.

ASSESSABLE

This Certifies, Jan 9th 1874.

THAT Bailey Gatzert is entitled to One Shares in the Capital Stock of the

Seattle and Walla Walla Railroad and Transportation Co.

TRANSFERABLE ON THE BOOKS OF THE COMPANY BY ENDORSEMENT HEREON AND SURRENDER OF THIS CERTIFICATE.

Roswell Scott, Secretary. (Dispatch Print, Seattle.) A. A. Denny President.

$100 EACH.

FIG. 3.1. SEATTLE AND WALLA WALLA STOCK CERTIFICATE. The owner of this stock certificate was Bailey Gatzert, mayor of Seattle when Joe Surber began to drive pilings into the tideflats for the S&WW trestle. Courtesy University of Washington Special Collections, UW 36354.

and across sagebrush plains. With a population of thirty-five hundred, Walla Walla was the largest city and county seat of the most populous and wealthiest county in Washington Territory. To fund the new enterprise, $10 million worth of stock was approved, selling for a hundred dollars a share, though anyone could offer up what they could, including labor, tools, and supplies, in return for stock.

So passionately did the city council believe in the importance of local control that they gave all the tideflats to the Seattle and Walla Walla. They did this through the city's forty-fourth ordinance, which bestowed on the railroad "all the tide-flats south of King street, in, under, around and about Elliott's Bay, from extreme high tide to extreme low tide and to deep water."[6] This was an incredible coup for the little railroad and its local owners, who would now be able to dictate the development of the area. They would be able to lay out routes across the muddy expanse, set the timetable for action, and control who else could do the same.

With the surveying completed over the winter of 1873–74, plans came together for a community workday to begin building the rail line to the east on Friday, May 1.[7] The historic day began at 5 A.M. with a brass band playing, cannons blasting, and church bells ringing. Early risers rode horses, walked, or traveled by wagon three miles or so to the meeting point, near where I now stand, east of the original mouth of the Duwamish. In 1874, the main route they would have taken south from the urban core was a narrow dirt-and-plank road that skirted the western edge of Beacon Hill. Because of slides and tides, it was typically passable only in the summer; the winter route was over the hill. Two boats also ferried people and supplies, but they got stuck in the mudflats south of their departure point, Atkin's Wharf, at the present-day intersection of First Avenue South and South King Street. The boats didn't move until the afternoon tide came in, though the would-be workers didn't suffer, since the supplies included enough liquor to disable a few travelers.

Town was deserted by nine o'clock, with a thousand men, women, and children working on the railroad line. "Never, perhaps, were before seen in any gang of railroad laborers, so many soft hands, white shirts and gold chains," observed one reporter.[8] After a fine lunch, dignitaries gave earnest speeches till local curmudgeon Henry Yesler uttered it was "time to quit fooling and go to work." (The following week, business in Seattle would be noticeably less rushed than usual as the workers recovered from their atypical exertions.)

By the end of the first day, Seattle's citizens had cleared, grubbed, and graded nearly a mile of land for their new railroad. The work, according to "experienced engineers," was equal to at least a thousand dollars in contract labor.[9] After the first day's enthusiasm, spirits lagged, with many fewer showing up the following week. But those who believed, or had been paid to do so, kept at the task, clearing and grading twelve miles up the Duwamish River valley by the end of October.

◆ ◆ ◆

Tracklaying south of the May Day picnic site had been relatively easy. Workers could access the area by road. They also benefited from the Duwamish River, which, as it meandered across its broad valley, had reworked its sediments and the lahar deposits, forming a flat, generally level surface, well suited for rail. The most challenging part was cutting down the forest of cedars, cottonwoods, and maples, some with diameters more than seven feet across.

Going north would be a different story, for instead of land, the track builders had to cross a protean stretch of real estate: the tidal flats, or delta, of the Duwamish River, which formed where the river met Elliott Bay, slowed down, and dropped its sediments. Looking west and north at low tide from the May Day work site, the Seattleites would have seen a two-and-a-half-mile-wide expanse of mud. The water, however, wouldn't have completely disappeared at low tide. One large channel and dozens of small channels would have been incised into the mud, branching and connecting in a complex dendritic pattern. The channels would have teemed with fish, waterfowl, and invertebrates such as shore crabs.

In addition to hearing the noise of the birds, the struggling men and women of Seattle would have noticed a distinct aroma wafting from the mudflats. Anaerobic bacteria buried in the ooze ate a rich brew of dead organic matter, then released pungent gases, which gave the mudflats an unfavorable reputation. Anyone working on the tracks in the warmer months would have perceived an even more redolent smell, since the gases spread farther and faster with increased temperatures. (Although the mudflats are gone now, the aroma still seeps out and wafts up to Beacon Hill, says a resident of the hill. She has also detected whiffs of doughnut grease, roasting coffee, and an acrid, pesticide-like smell.)

Workers could also have looked toward the river's mouth and seen two low islands covered in grasses. The islands blocked a view of three additional islands, the largest about 160 acres; a remnant of that is now known as Kellogg Island. The islands remained above the water at all but the highest tides.

During a typical workday, the Duwamish tidal flats slowly vanished. The rising tide first would have covered the northern edge of the mud, which ended at a line running roughly eastward from Salty's restaurant on the east side of West Seattle, across the northern end of Harbor Island, and curving up to intersect CenturyLink Field. Water then would have filled in the incised channels and, by high tide, covered the entire tidal flats from Beacon Hill to West Seattle, in places up to twenty feet deep. The sea would have reclaimed its territory.

This unpredictable characteristic made the tideflats problematic. How could one own such a place? Did you own land or water, and could you legally designate such an indefinable location, especially if no one could survey it? Addressing ownership issues regarding these twenty-two-hundred-plus acres would become one of the epic conflicts in the city's early history.

If you want to get a feel for what the original tideflats looked like, you are in luck. Although the Duwamish River has been channelized since 1914, efforts have been made to restore some of the natural habitat, particularly near Kellogg Island and at an eight-acre wetland located between State Route 509 and Second Avenue Southwest, just south of Highland Park Way Southwest. None of these spots are all that exciting, but if you have the time and patience to watch the tide change, you will be rewarded with an insight into the incredible dynamic and metamorphosis of the great tideflats that were once a dominant part of the life of Seattle.[10]

What these locations won't show, however, is the importance of the tideflats to the Duwamish people who lived here. Although archaeologists do not have extensive evidence of how Native people used the tideflats—because of the dynamic nature of the river, including the fact that it didn't reach its present mouth until about a thousand years ago—the midden site at Terminal 107 Park reveals that cockles, bent-nose clams, and littleneck clams were collected, as well as species found in rockier habitats, such as mussels and barnacles. The midden also contained the bones of harbor seals, which could have traveled several miles up the Duwamish. Fishing weirs found upstream would have facilitated the catching of migratory salmon, one of the most important food sources for those who lived along the river.

"It was part of our food source, our grocery store," says Warren King George of the Muckleshoot Tribe, who retain treaty rights to hunt, gather roots and berries, and fish in what are known as their "usual and accustomed grounds and stations" on the Duwamish River. Barnacles, clams, and mussels; seaweed, eelgrass, and kelp; and salmon, herring, and flounder were some of the traditional foods harvested from the tideflats. "People don't understand the bounty of the land precontact. There's no modern scale for understanding it," says King George. "Losing it was really heartbreaking."[11]

But there is some hope on the former tideflats. On one of my explorations, I noticed a pile of sticks wedged into the top of a cell-phone tower near the base of Beacon Hill. Looking more closely, I could see a gray bird standing on the tower near the nest. It was an osprey, a bird formerly known as the sea hawk and one never found far from water, which means never far from salmon. If an osprey can survive on these made lands of concrete, perhaps it is a sign that despite what was done to the tideflats, the nature of this place has not been totally obliterated.

Osprey also exemplify our changing attitudes to the natural world. Where once we called birds such as osprey, eagles, and hawks, varmints and tried to eliminate them, we now embrace them. We let osprey build nests on cell towers, and we erect nest platforms for them. We designate their main food source as an endangered species and do what we can to protect it. We've even gone so far as to name a sports team after the osprey. We are realizing that in order for us to live here, we need to learn to live with osprey and not to control them. It is a lesson that we are also beginning to learn in relation to how we live with the landscape we have.

◆ ◆ ◆

Work on the Seattle and Walla Walla stopped in late 1874. The national economic panic of 1873, and Seattle's still-meager existence, translated into an insufficient number of citizens responding to the Seattle Spirit and putting up the cash necessary to fund tracklaying. But then, in 1876, a savior appeared in the guise of Scotsman James Colman. After living in San Francisco, Colman had made a fortune leasing and operating Henry Yesler's mill. He pledged twenty thousand dollars to get the seemingly left-for-dead railroad up and running. Either embarrassed or enticed by Colman, many in Seattle followed suit and work began again.

The early settlers had a major reason to make sure that the S&WW succeeded. The original core of Seattle had little room for a railroad. Because of the rapid growth of the urban infrastructure, all the terrain where rail could have come into the business center was taken with roads and buildings. And even if no infrastructure had been built, the hills and ridges would have restricted access to the business district. The only possible routes skirted the city, and the only way to build a track there was to cross water, either at the base of the steep bluffs that rimmed Elliott Bay or on the tideflats.

Without rail, Seattle could not hope to expand as a city. It needed train service not only to bring people to the young and growing town but also for moving resources to and from the city. Nowhere was this more apparent than with coal. In 1853, deposits later described as richer than "any mine yet worked on the Pacific Coast of America" had been discovered east of Lake Washington, around Newcastle and Renton.[12] The landscape of hills, lake, and two winding rivers, however, had long prevented the full exploitation of the eastside coal beds. In 1864, for example, transportation to and from the coal beds took twenty days.

Little had improved by 1876. One route involved travel by barge across Lake Washington, out its mouth at the Black River to the Duwamish River, and down to Elliott Bay. Because of snags and sandbars on the Duwamish, the journey could take up to a week. On the "faster" route, the coal was loaded on railroad cars at the mine and lowered nine hundred feet by tram to Lake Washington, where the cars traveled on a barge to Union Bay, or what was also known at the time as Foster's Bay. After another tram trip, across the narrow neck of land now crossed by State Route 520, the coal cars were placed on a second barge for a two-mile journey across Lake Union to the south end of the lake and a waiting train that carried the coal its final mile to a final tram, which lowered the ore down to a massive coal bunker at the base of Pike Street.[13] When that train, aptly named the *Ant*, had made its initial run over one mile of track on March 27, 1872, the town celebrated the opening of its first railroad, the Seattle Coal and Transportation Company, with "unanimity never before equaled on any occasion in the city."[14]

Building the S&WW meant that coal could be loaded on a single train at the Newcastle mines, travel through Renton to the Duwamish River valley, and remain on the train until it reached tidewater in Seattle a few hours later. The final destination would be the insatiable market of San Francisco, where coal was a cheap and essential fuel for industry and transportation. San Francisco in 1870 was the tenth-largest city in the country, with 149,473 people, but neither California nor Oregon had the vast coal reserves necessary to fuel San Francisco's economy.

◆ ◆ ◆

Crews began their attempt to cross the tideflats on May 11, 1876, at Atkin's Wharf, which jutted out onto the flats and was the southernmost point in Seattle. Former chief of police Joe Surber floated a scow up to the dock and begin to drive logs into the Duwamish mud with a pile driver. A rather simple machine, with technology dating back to at least the heyday of the Romans, a pile driver operated on the hammer principle, using a heavy iron weight to pound a wooden pole into the

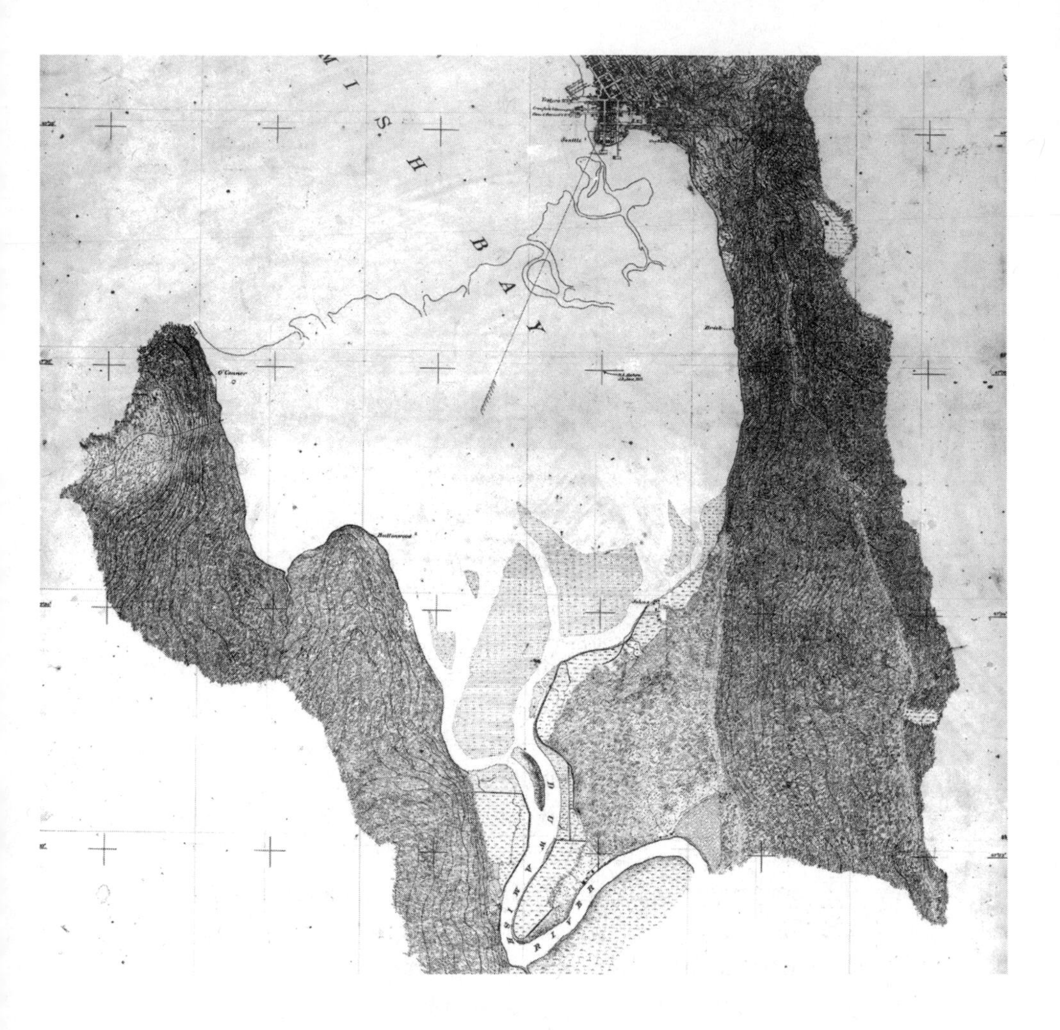

FIG. 3.2. DETAIL OF THE U.S. COAST SURVEY MAP OF DU WAMISH BAY, 1875. Twenty years or so after Seattle's settlement, little new land in the city had been made. The tidal swamp behind Maynard Point, however, had almost been filled in. Yesler's Wharf had grown substantially since the Phelps map was created, and in this map it is the largest structure jutting out into Elliott Bay. On the left side of the southern end of the point is Atkin's Wharf, where Joe Surber started to drive pilings in the tideflats for the Seattle and Walla Walla Railroad in May 1876. Library of Congress, Geography and Map Division.

ground. We have no way of knowing the number, but pile drivers must have pounded in hundreds of thousands of piles across the city. Without a doubt, few machines contributed more to the transformation of the Seattle landscape.

Surber's pile driver sat on a bargelike boat that was about equal in length to the fifty-foot-tall tower that supported the iron hammer. A shack took up the back half of the boat. Inside was the steam engine that drove the hammer. Surber most likely used Douglas fir for his pilings; the day before he started, a steamer had arrived with a boom of logs from Port Blakely on Bainbridge Island. With mud more than thirty feet deep on the tideflats, Surber needed piles up to sixty-five feet long. Two whacks of the three-thousand-pound hammer would push the piles down through the soft mud to harder material below; more blows—sometimes over 150—were required to sink the log to a stable depth.[15] No record exists of how fast Surber worked, but a report in the 1890s described pile drivers putting in from twenty-six to two hundred piles per day.[16]

Although it might seem logical to drive bare piles, that did not always occur around Puget Sound. Piles often had bark on them. During the growing season, workers girdled the bark just above the base, which killed the tree, and then let the bark shrink for several months before cutting the tree. They did so because of the "terror of the dock builder," a small bivalve known as the teredo, or shipworm.[17] Shipworms resemble a worm sporting a conical helmet—the bivalved shell—which the animal uses to drill burrows into wood submerged in salt water. So many shipworms can infest a piling that it makes Swiss cheese look solid; their drilling could render a floating log worthless in a month and destroy pilings in a year. To combat the destruction, lumber mills soaked logs in creosote, coal tar, asbestos, castor oil, and strychnine, but the poisons apparently "pleased the teredo much, seeming to act as a condiment on what must have been a rather monotonous bill of fare."[18] Bark slowed down the shipworms, but only until it decayed or fell off.

By late August, Surber had driven a line of piles across the tideflats. Workers then began to lay supports across the trestle, followed by the narrow-gauge rail. When finished, the trestle rose several feet above the highest tides. The *P-I* reported that it quickly became a popular place for couples to go on Sunday afternoon promenades, though they had to watch for trains, which began to compete with the lovers in late October.

Train service on the Seattle and Walla Walla began officially on March 7, 1877. Two days later, the first coal arrived from Newcastle. With the completion of the S&WW, coal exports from Seattle to San Francisco jumped from 4,918 tons in 1871 to 132,263 tons in 1879; and by 1883, Seattle was providing 22 percent of the coal produced on the Pacific Coast.

Although we usually don't consider it today, coal helped make Seattle the most important town in Puget Sound.[19] With coal, the young community had a second economic driver beyond what every town in the Pacific Northwest had: that is, they all possessed a seemingly endless supply of first-rate lumber. Not only did Seattle gain financially by selling coal and by having coal-related jobs, but it also supplied support services to the developing coal towns, such as goods for workers living at the mines, lumber for buildings, tool shops to build and repair machinery, and a foundry to make rail cars. Coal by itself did not make Seattle, but having an additional natural resource to exploit made the city more than a timber town. Coal remained an important export commodity till around 1910, when petroleum started to be more widely used.

In building that first trestle, Joe Surber proved that there was a way to bridge the mud and flood of the tideflats. It was the first layer of scaffolding that would ultimately lead to the filling in of the tideflats. It was, in the words of an editorialist writing in the *P-I*, an "object of vital importance to us as a community."[20] Seattleites showed that they could cross the vast tidal flats, making the city open to one of its ultimate goals—connection with a transcontinental rail. The S&WW would not be the link, however; it never went past the Renton coalfields. (Prophetically, the railroad did run the route still followed by trains entering Seattle today. And the location of the May 1, 1874, picnic is very close to where the city's modern train yards are located.)

Just as I wondered about the decision to settle around Maynard Point, I have to wonder what was going on in the minds of the supporters of the Seattle and Walla Walla. Were they so optimistic and hubristic that they thought they could build a railroad by themselves across the state to Walla Walla? Did they have any clue what it would truly take to accomplish their goal, or were they so savvy that they knew their attempt would attract the attention of people who could do it and would perhaps buy the railroad from them?

I tend to believe that it was a combination of all of the above. Seattleites felt they had been backed up against a wall and jilted by the Northern Pacific, and they certainly weren't going to let lowly Tacoma triumph at Seattle's expense. Building across the tideflats was what they *had* to do; in their minds, they had no other decision. Seattle would not survive without that railroad. Because they lived in an era when men thought they could triumph over nature, Seattleites believed they could move forward as if the natural rhythms of earth and sea did not exist. Who cared if the tides came in twice a day, when you could drive a forest of logs into the ground and develop a platform for industry? In early Seattle, apparently few did. It would take until the twenty-first century for people to care, to try to respond accordingly.

◆ ◆ ◆

To get a feel for how Surber's trestle transformed topography, I ride to Jose Rizal Park, on the northwest corner of Beacon Hill. From this little-known park, named for a Filipino social reformer, I can look out to what would have been Maynard Point and the tideflats that once stretched from a point below me, at the base of Beacon Hill, past what is now CenturyLink Field and Safeco Field to what is now the eastern edge of West Seattle. Two and a half miles south of the baseball stadium would have been the mouth of the Duwamish River. Most everything west of Jose Rizal Park, including the two stadiums, all of the Port of Seattle's massive cranes, Harbor Island, and the Interstate 90 off-ramps, sits on the made land of the former tideflats.

The best way to illustrate the evolution of the tideflats is to turn to one of my favorite pictures of Seattle. Taken by an unknown photographer in 1884, from roughly where I stand, it shows how quickly Seattleites began to change this section of the city once they knew they could. Amazingly, Joe Surber's Seattle and Walla Walla train tracks are no more. Not only was the original trestle no more, but also gone was the Seattle and Walla Walla. Colman and his co-owners sold the railroad for $350,000 in October 1880 to Henry Villard, who would also purchase the Northern Pacific Railway. Once again, it looks as if the brash Seattleites knew what they were doing. In this photograph, most of Surber's pilings still stand in a line cutting across the tideflats; but within a little over a year after completion, shipworms had so weakened the structure that the railroad had to abandon it. On his second try, Surber pounded his piles for a trestle closer to Beacon Hill, more or

FIG. 3.3. VIEW OF DUWAMISH RIVER TIDEFLATS FROM BEACON HILL, 1884. The wide white trestle that crosses the tideflats carried the Puget Sound Shore Railway, a subsidiary of the Northern Pacific. Next to it lies the abandoned trestle of the Seattle and Walla Walla. Closest to shore and curving toward the massive coal bunkers in Elliott Bay is the Puget Sound and Columbia Railroad, another subsidiary of the Northern Pacific. Note the abundant rafts of logs floating around the sawmills at what is the southernmost point of the city. Courtesy Museum of History and Industry, image SHS60237.

less following the path of our most important transportation route in modern times, Interstate 5.

More tangible evidence of Seattle's southward growth across the tideflats is the cluster of buildings erected on pilings south of the big coal bunker (next to the many-masted ship) in the 1884 photograph. They include a furniture manufacturer, a sash and blind company, and several sawmills, in particular the huge new Stetson and Post facility, which had begun in 1875 as a grist mill and grown to become one of the largest mills on Puget Sound. In the early 1880s, the lumber yard employed more than a hundred men, who cut 18 million feet of lumber and 4 million laths in 1884.[21] In contrast to other mills around Puget Sound, Stetson and Post didn't export most of its sawn timber. Instead, because of the rapid growth of the city, most of the wood stayed in Seattle and was used to construct businesses and apartments and the one thousand homes built in 1883.[22] This would have significant consequences in 1889.

Mills such as Stetson and Post played a key role in the development of the city's south-end topography. Every cut log, lath, sash, blind, shingle, molding, and door produced sawdust, and the easiest way to dispose of it was simply to dump it on the tideflats, along with all the mistakes, detritus, and other debris. Eventually enough material accumulated that the tide could not cover it, and new land was born. As Bostonians had started doing in 1641, Seattleites were wharfing out their town. You can see this in the middle right-hand, or northern, portion of the 1884 photograph. The trestle that runs due east-west is the Jackson Street Bridge. Water comes up to the base of the bridge; north of it is made land created from sawdust and trash, along with dirt from regrading work done by Front Street regrader George Edwards. And so the city grew, block by block, expanding south on waves of waste.

As the lumber yards illustrate, the primary purpose of the newly made land was industry. It was an ideal location: flat, isolated, easy to expand (just dump your waste on the tideflats), and necessary. Seattle had never really faced a shortage of space for residences. They could be built on most of the hills, which were often the fashionable place to live, as they were in cities across the country. In contrast, most industries, and particularly ones in a city based on natural-resource extraction, needed the new land because of their reliance on access to outside transportation. This can be seen in the 1884 photograph in structures such as the coal bunker, the docks on Elliott Bay, and of course, the railroad tracks.

Yet local entrepreneurs did not necessarily jump at the opportunity to take advantage of the new land. An article in the *P-I* reports that "nothing with them could be more repugnant" than opening businesses on fill.[23] One reason most likely was the aroma, which was notoriously unpleasant becasue of the rotting garbage and waste of the fill, but another, notes historian Paul Dorpat, was racism. As so often still happens, newcomers, in this case Chinese workers who labored on the railroads and regrading projects, and who were essential for the city's growth, ended up living on low-quality land; it was cheap and unclaimed and located away from the more established upland homes and businesses of Seattle's more prominent citizens. Living on made land did not, however, create a refuge for Chinese residents.

The first Chinese man had arrived in Seattle in 1860, having perhaps, like many of his countrymen, initially landed in San Francisco following the gold rush. By 1880, Chinese workers made up 13 percent of the city's labor force over the age of eighteen. Because of societal pressures, most lived within a block or two of the new land expanding into the former marsh around Washington Street and Second Street, where they worked as grocers, sold dry goods, and ran laundries. When the nationwide economic downturn of the mid-1880s hit Seattle, labor and civic leaders scapegoated Chinese workers as the source of the city's problems, which ultimately led to one of Seattle's most horrific events, the forced expulsion of most of the Chinese residents in 1885–86.

The low-quality property being made during this wave of land creation also provided a refuge for Native peoples. Three or four blocks west from where most of the Chinese workers had resided was a dab of rock and rubble known as Ballast Island. As the name implies, the island was born of the ballast merchant ships dumped after arrival in Seattle. Historically, trade vessels needed to carry dense but inexpensive ballast, such as rocks or bricks, which would be dumped and replaced with cargo at a port of call. Since San Francisco was the city's earliest trading partner, rock from there became the original source of ballast that ended up in Seattle. One ship dumped as much as three hundred tons of rock and sand, from quarries on Telegraph Hill, into the water south of Yesler's Wharf. Expanding trade led to rock from around the world building up the island. One source described finding material there from Valparaiso to Sydney to Boston to Liverpool.[24]

Ballast Island soon grew large enough to show up on maps and in photographs, with the latter typically featuring canoes and tents on the island. Ironically, the artificial island made of exotic rocks was one of the few spots in Seattle in the late 1880s and early 1890s where Native people were tolerated, notes Coll Thrush in *Native Seattle*. In addition to being a stopping-over point for tribes from outside the region who were headed to work at the Kent and Auburn hop fields, Ballast became a refuge for locals, including several Native families that white settlers had burned out of their homes in West Seattle. In Thrush's sobering language, Ballast Island exemplified how "urban development and Indian dispossession went hand in hand."[25]

The city's marginalized land provided space for a third group of people typically ostracized by society. California businessman John Pinnell opened the town's first official brothel on the property that had been filled by Yesler's wheelbarrow man, Dutch Ned; Pinnell's ladies were known as "Sawdust women." By the late 1800s, the district had grown substantially, earning its well-known names of Skid Road and Lava Beds, a reference to the incendiary nature of both its denizens and the sawdust they lived upon.[26] During the Depression, a new group of disenfranchised people established a Hooverville in the area of what is now Pier 46, due west of CenturyLink. Hundreds of people put up wood-and-tin shacks on the land, where decades-old pilings still poked up through the fill. When very high tides came, rising water made islands out of some of the shacks built closest to Elliott Bay.[27]

◆ ◆ ◆

Although the Seattle Spirit had certainly driven the developers of the Seattle and Walla Walla and subsequent railroads to cross the tideflats, not everyone bought into the idea of lifting up Seattle and helping others. For many, simple greed was a far stronger motivation, as is evident in the 1884 photograph and its curious forests of pilings enclosing square spaces on the tideflats. They were put up by people known as jumpers and squatters. Squatters were the owners of shore property who built out into tidal areas that they interpreted, or hoped to claim, as their own. Jumpers simply drove in piles wherever and whenever they could. They, too, hoped that doing so would give them ownership of their little square of the tideflats.

Their attempt to control the tideflats was Seattle's version of the Wild West, where schemers and swindlers tried everything they could to acquire property and convert it to cash. One of the more audacious attempts, and a ploy tried across the West, was to use government scrip, which allowed the owner to place claims with the U.S. government on "unoccupied and unappropriated public lands." Between January and October 1889, more than a dozen people, from locals to outside agents, filed claims for more than eight hundred acres on the tideflats.[28]

A second enterprising group tried to pass themselves off as oyster cultivators. An 1877 territorial law allowed people to plant oysters in areas with no native oysters, stake out the property, and obtain a title to their mollusk farm. It was a great idea, except that earlier Seattle settlers had tried to raise oysters and quickly learned that the Duwamish's muddy waters killed their progeny. Still, the optimistic, or opportunistic, tried this ruse as a means to acquire roughly six hundred acres of tideflats.[29]

Those with more moxie took the direct route, particularly after January 1888. In that month Henry Yesler decided to expand his sawmill farther out onto the tideflats. Problem was, Henry didn't own that property. The area was part of the land given to the Seattle and Walla Walla Railroad and Transportation Company in 1873 and was now owned by Henry Villard and the Northern Pacific. When the railroad company decided not to contest Yesler's usurpation, jumpers and squatters took it as an unwritten invitation to lay claims for themselves.

The Seattle establishment could hardly have foreseen that their impassioned 1873 donation of the tideflats to the S&WW, which essentially meant to themselves, would result fifteen years later in hordes of opportunists seeking to take over the territory south of downtown. In one week, the usurpers drove more than two thousand piles into the flats; Joe Surber told a reporter for the *P-I* that he had orders for an additional one thousand piles. One zealous claimant hired twenty carpenters to prefabricate cabins, which they floated out late one night piece by piece. The next day, when low tide exposed the flats, the men erected three cabins, each measuring twenty by twenty-four feet, in less than two hours. Apparently, the head of the project had good taste, since one newspaper noted that the cabins had "matched floors, doors, and windows."[30] In order to protect their claims, other squatters built their own shacks (though no one mentions their decor) and hired armed guards. The only incident occurred when one impetuous attorney accidentally blasted a hole in his shack with his pistol.

Such unseemly actions were noted with derision by the more august members of Seattle society. None were more contemptuous than prominent Seattle attorney Thomas Burke, one of the more intriguing early Seattleites. He had arrived in 1875 with a law degree and ten dollars in his pocket; but he had quickly established himself as a tenacious attorney and gifted orator and betrothed himself to the daughter of John McGilvra, a well-connected and wealthy lawyer. Standing a bit over five feet tall, Burke had a bit of a Napoleon complex: arrogant and feisty, he was always ready to fight for what he believed in, which often meant Seattle. Like many of his era, Burke was a fierce proponent of the Seattle Spirit, though he put a more personal twist on it than most. He would do whatever necessary to protect and promote the city's interests, but he would also intertwine his own interests generally, for personal financial gain.

Writing to a friend, Burke noted, "Here in Seattle the craze for salt water has broken out again with greater virulence than ever before. A swarm of salt water lunatics, of high and low degree, have alighted, like so many cawing crows on the mudflats in the Southern part of town, and have made the place a veritable sight with piles stuck all over it."[31] Unfortunately, he added, the charitable sea had failed to drown the knaves. Part of Burke's consternation arose because the "salt water crew" had the audacity to "plant themselves on his property," which he had paid for in gold. Although he might lambaste his lessers, Burke did not oppose land grabbing.

On April 15, 1885, he and Daniel Gilman had led a group of Seattle investors in forming yet another railroad. The Seattle Lake Shore and Eastern Railway Company would achieve the long-awaited goal of laying track across the Cascade mountains, from Seattle to eastern Washington, where it would link to a Union Pacific Railroad spur. Instead of heading south, the SLS&E would go north from downtown, through Smith's Cove (the present-day Interbay), past Salmon Bay and property owned by Burke, and finally around the north ends of Lake Union and Lake Washington.

Burke and Gilman—a paired set of names now best known in relation to the eponymous trail that has replaced the tracks—chose this route for two reasons. It made sense topographically. Once the train made it past the bluffs north of downtown, it would have a relatively level route along the northern shore of Lake Washington and access to the little communities starting to develop north of the city—which had located there because of the more level terrain. (This level land is part of a curious geologic enigma, the northwest-southeast trending diagonal slash of lowland where the Lake Washington Ship Canal is.) Coal, too, influenced the choice of route, because trains would pass by coalfields near Squak Lake (now Lake Sammamish) and iron mines near Snoqualmie Pass owned by Burke and Arthur Denny.[32] With coal and iron, the Seattle area would soon become the "Pittsburgh of the Pacific Coast," gushed the *P-I*.[33]

To facilitate construction of this northern route, Burke orchestrated one of the great land grabs in Seattle history, though technically the deal involved no land that then existed. He had to do so because

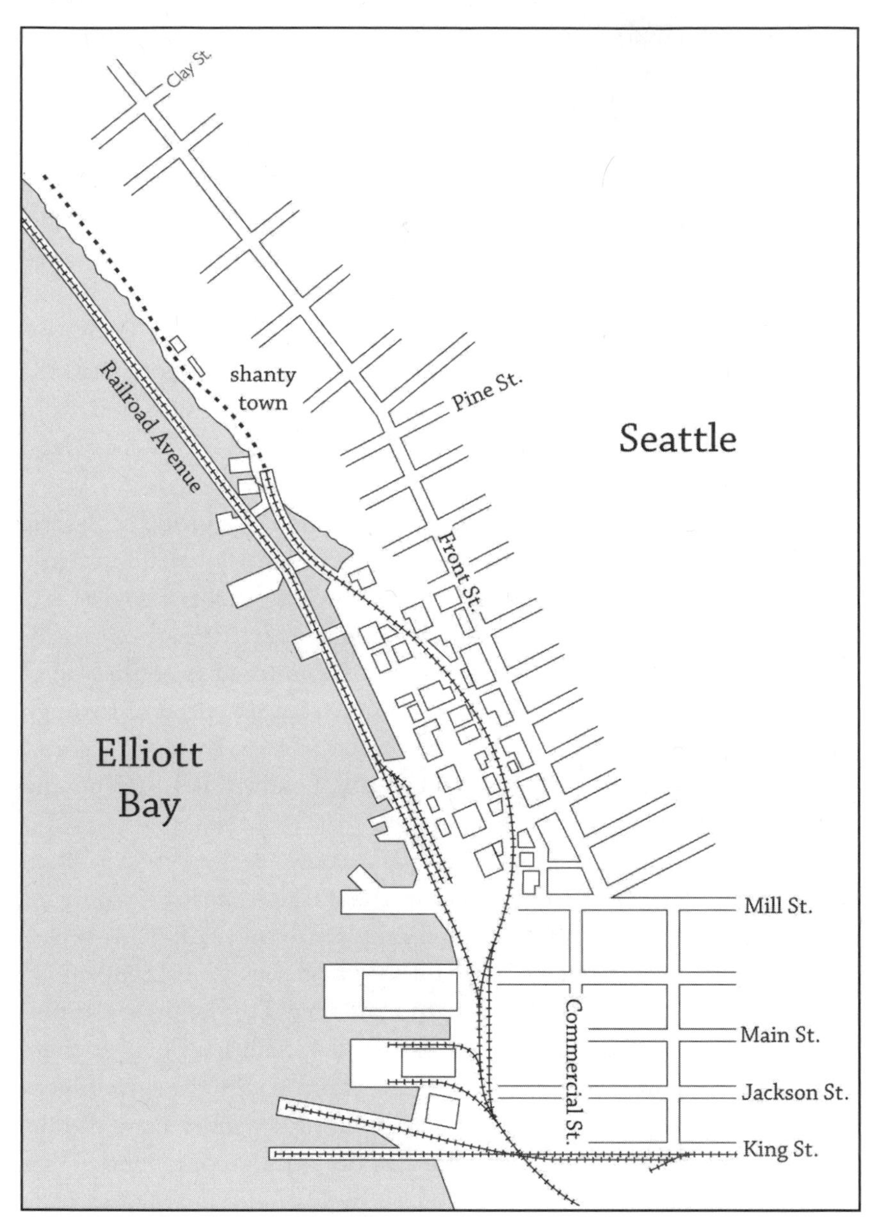

FIG. 3.4. RAM'S HORN AND SEATTLE LAKE SHORE AND EASTERN RAILWAY. This map shows Seattle in 1888 and includes the historic names of streets. The Ram's Horn is the railroad track that curves inland; the SLS&E runs on a trestle along what is labeled as Railroad Avenue. Courtesy Washington State University Press, from Kurt Armbruster, *Orphan Road: The Railroad Comes to Seattle, 1853–1911* (Pullman: Washington State University, 1999).

FIG. 3.5. SEATTLE LAKE SHORE AND EASTERN RAILWAY TRESTLE, 1887–88. Charles Morford's photograph of the trestle, which Joe Surber built, includes, to the right, the area that was then just becoming known as Belltown. Note the round building in the upper right with the triangular roof. This was Dr. Orlando Root's octagon house, discussed in chapter 5. Courtesy Paul Dorpat and William Mix.

Henry Villard, too, understood the topographic importance of controlling train access along the northern shoreline of Elliott Bay. In 1882, the city council had granted Villard a right-of-way along the waterfront from King Street to Clay Street, which soon acquired the name "Ram's Horn" because of its curving route through public and private property. In order to circumvent the Ram's Horn, Burke convinced the city council to grant his railroad the perpetual right to lay tracks on a 120-foot-wide street west of the Ram's Horn.[34] Railroad Avenue, as the new entity would be called, would not actually be on land; it would have to be built on trestles over Elliott Bay from Mill Street to Smith's Cove.

Work on the initial Railroad Avenue trestle began in early March 1887. Joe Surber won the contract to build 16,640 linear feet of trestle along the waterfront. To complete the job in the required three months, Surber acquired three pile drivers and had one specially built for the project. The wood for the six thousand piles came from forests around Lake Washington and Henderson Bay (just east of Gig Harbor).[35] In an attempt to combat shipworms, the piles would be covered in creosote.

By August 1887, Burke was able to lead his fellow railroad company owners out to pound the first spike on the trestle, and by October 25 the SLS&E was running twice-a-day trains to Union Bay, with stops at Ballard, Ross, Fremont, Edgewater, Latona, Brooklyn, Ravenna, and Yesler Village. Round-trip was about ninety minutes. To go farther, passengers had to take a steamer, which stopped at towns on Lake Washington and Lake Sammamish. For Thanksgiving, the SLS&E celebrated with service to the end of the line, near present-day Bothell.

◆ ◆ ◆

Sometime after workers finished erecting the SLS&E trestle, a young man named Charles Morford ventured out onto the waterfront tracks and took a stunning photograph. Morford pointed his camera north from a spot that is now directly under the Alaskan Way Viaduct, at Seneca Street, but which in 1888 was about ten feet above Elliott Bay. The trestle and tracks emerge from the southwest corner of the image, head north over water, slice between two warehouses of a salmon cannery, and curve along the steep bluffs that I described previously while standing at Lenora Street. The track and trestle fade off into the distance, heading toward Smith's Cove, a stark line of uniformity against what looks like the recently, and only lightly, logged slopes of present-day Queen Anne Hill and Magnolia.

Four features make the photograph compelling. First, consider the buildings on the slope above Elliott Bay. Most are small homes, or what the Sanborn Map and Publishing Company's 1888 Fire Insurance Map for Seattle labels as "Squatters Shanties."[36] Below these shanties are more residences, the cabins and shacks around the Bell Street Ravine, which, after the construction of the trestle, were in an even less desirable position, towered over by a bluff to the east that could slide at any time and hemmed in on the west by an ever-expanding rail system. Not until reaching the top of the slope does one encounter the more substantial buildings of Seattle's wealthier residents. Morford's photograph provides a good illustration of the topographic divide in central Seattle, with the marginalized citizens generally restricted to the least favorable terrain.

The second feature of interest is the amount of wood used in the trestles. All of the millions of board feet of timber had come from the great old-growth forests that covered the region, which meant that it was strong, solid, high-quality timber, perfect for supporting a train system built where trains normally didn't run, over water. Part of the reason Seattle could build such a trestle system was the local lumber industry. Wood was cheap and seemed inexhaustible. In 1888, mills around Puget Sound produced 455 million board feet of wood, of which Seattle mills contributed 100 million feet. Not only was the SLS&E providing access to the world, it was providing jobs.

The third feature is the water under and to the east of the trestle. Morford took his photo at a good distance from land. The Ram's Horn runs right along the edge of the bluffs, but the SLS&E at this point is nearly 300 feet from the 1888 shoreline, and an additional 150 feet or so from the original shoreline.[37] Like the first trestle across the Duwamish, the new SLS&E trestle is a manifest sign of the city's destiny to grow. We can already see this in Morford's photograph, where a plank-covered pier has reached out from the shoreline to connect to the train tracks. Soon, more piers and wharves and warehouses will spread over the water and then extend farther into the bay; one of Burke's sly moves was to ensure that his railroad company retained exclusive right to the thirty feet of Railroad Avenue closest to land. Any other railroad company that wanted a rail line running north along the waterfront would have to build a trestle and lay rail west of Burke's tracks, paralleling them, in deeper water. This led to more difficulties in construction and made access to warehouses on land more challenging: side tracks from later rail lines had to cross over Burke's to reach the shoreline warehouses. If they got in his way, he wrote, "we shall be able to bring powerful pressure to bear on them to make them behave themselves."[38]

The final captivating aspect of Morford's photograph is the gentle curve of the Ram's Horn railroad trestle, which bends from shore at a diagonal toward the SLS&E trestle; they almost touch a bit north of what is now University Street. Two railroads that run next to each other. Two railroads that fought tooth and nail for decades to find a way into and through Seattle. In this modern age, it's hard to fathom how much effort was expended to make railroads work in Seattle; but their builders had no other choice, because the route along the waterfront was the only viable way north. Kurt Armbruster, author of the *Orphan Road: The Railroad Comes to Seattle, 1853–1911*, says that this section of road along the waterfront was arguably the most contentious piece of rail in the city.

Ironically, the importance of these local rail lines lay not in connecting Seattle to the rest of the nation but in tying the young city to the rest of the region. Totaling less than 150 miles of track, the local lines made Seattle essential to its nearby neighbors by receiving their raw goods, such as coal, poultry, grain, and vegetables, and providing their basic supplies, including beer, rubber boots, fishnets, and donkey engines (steam-powered winches). Even after the transcontinental railroads arrived, rail shipments within the state dwarfed those out of state. The local lines were, argued one economic historian, far more important than transcontinental railroads in explaining why Seattle became the major city of the state, and his argument had nothing to do with the unquantifiable effects, such as their facilitation of the physical growth of Seattle.[39]

◆ ◆ ◆

Before anyone could build too many piers out to the SLS&E trestle, however, 120 acres of the city's main business district burned in the Great Fire of 1889. The inferno did not limit itself to the land but also consumed Seattle's piers, wharves, and trestles, leaving behind spaghettied train tracks, a grid of decapitated pilings, and a wasteland of blackened, splintered, and shattered infrastructure. In the haunting photographs taken in the days following the fire, it looks as if some malevolent force had peeled off the ersatz skin of the waterfront and revealed the fragile framework hidden beneath. And just like that, within a day, years of work spent on expanding and improving the city was rendered useless by what the *P-I* called the "seething mass of crackling flames."[40]

Once again the mighty Seattle Spirit arose and Seattleites rallied to rebuild. One advantage of the fire was that it created tons upon tons of ruined buildings perfect for dumping into the bay, extending streets, and making new land. So much material ended up on the tideflats and in Elliott Bay that anglers who used to congregate on the downtown wharves now had to head across the bay to West Seattle to find clear water favorable to fish.[41] Within weeks of the Great Fire, at least twenty pile drivers were at work; but even though all the "old and discarded pile

FIG. 3.6. LOOKING WEST FROM COLUMBIA STREET AND SECOND AVENUE (?) AFTER THE GREAT FIRE, JUNE 6, 1889. The devastation wrought by Seattle's Great Fire is evident in this view looking west down Columbia Street. Three pile drivers, with their towering frames, are at work along the waterfront. The forest of pilings reveals where and how much land has been made along the waterfront: anyplace pilings stand was originally underwater at high tide. Courtesy University of Washington Special Collections, UW 29369z.

drivers in town had been hunted up [and] hastily repaired," they could not meet the needs of builders.[42] Every pier was being built bigger and extended farther out into the water. Railroad Avenue grew, too, as the city drew up new ordinances to make it more substantial.

By the early 1900s, railroad companies had crammed nine parallel sets of train tracks onto the trestle system that extended out over Elliott Bay. This does not include transfer tracks and the tracks that curved

FIG. 3.7. RAILROAD AVENUE, CIRCA 1905. By 1905, the trestle system that supported the train tracks on Railroad Avenue was so well integrated that it looked like a road on land. It wasn't. All the tracks rested on trestles built over Elliott Bay, meaning that if someone fell through—and they did in places where there were gaps—they would end up in water. Within a few years, automobiles would also start to use this same space, creating a travel corridor noted for epic chaos. Courtesy University of Washington Special Collections, Hegg 2235.

over to warehouses onshore and on docks.[43] More than a hundred trains passed over the tracks every day. Not only did one have to cross paths with trains and risk getting hit, but those who wanted to reach the warehouses and wharves along the water also had to cross train tracks that had gaps on either side, as well as splintered, uneven, and sometimes absent, planking, and risked falling through—a daunting adventure not recommended for timid, slow, or clumsy pedestrians. At least ten people died in train accidents in one two-year period, with scores more maimed over the years.[44]

The tracks were so busy and dangerous that the problems they created led to Seattle's first attempt at a new method of bypassing the city's topography. Instead of making new land, the Great Northern Railroad Company, which had arrived in 1893, would dig a tunnel from the company's soon-to-be-built Union Depot (now called King Street Station) north to Railroad Avenue, at Virginia Street. Crews started digging in 1903 from either end, with the deepest section 140 feet below the surface. Like the modern State Route 99 tunnel, the Great Northern tunnel went through unconsolidated sediments deposited during the last ice age, which made it relatively easy for digging by hand. Both generations of tunnel proponents relied on the same justifications—better access, less noise, and safer travel. Although the 1903 crews had only picks, shovels, wheelbarrows, and muscle to dig their mile-long route, they finished on time, with the two crews meeting almost perfectly in line with each other. The Great Northern tunnel opened in 1905 and still provides rail access under the city.

In what could be a lesson for our times, the tunnel did not solve the problems its proponents hoped it would. Accidents continued to be common as the marginally restrained Railroad Avenue train traffic competed with traffic created by the newfangled automobile, as well as with bikes and pedestrians, all of which used Railroad Avenue as a transportation corridor. And heaven forbid if you fell through one of the many holes in the planks covering the trestles. A city report described the area under Railroad Avenue as a cesspool "filled with the most horrible filth," because businesses and individuals continued to expel their waste below the trestles.[45]

One proposed cleanup solution was to eliminate the space under the trestles by building a seawall on the western edge of Railroad Avenue and filling in the area between the wall and shoreline. An additional impetus in the plan to build a seawall was the expected 1914 opening of the Panama Canal, which would cut more than seven thousand miles off the trip by sea from New York to Seattle. Like every port on the West Coast, Seattle believed that it would benefit from the canal, but only if it adapted to new and larger ships, improved its rail connections to terminals, dredged deeper channels, and built new docks, warehouses, and wharves, all of which would be aided by the presence of a seawall.

But nothing was done till 1916, when a twenty-foot-tall concrete seawall, fronted with riprap and backfilled with dirt, was built between Washington and Madison Streets.[46] Another eighteen years would pass until the city, with funding from the federal government, completed the seawall to the north, using corrugated steel sheet piling, a wood platform, concrete slabs, and wood piles up to 110 feet long. Behind the wall, crews dumped 250,000 cubic yards of sand and gravel to create a surface for a new 150-foot-wide roadway for rails and automobile traffic.

We now know that the builders of that seawall and roadway did not make the best decisions, primarily because they did not know what modern engineers know. The fill behind the wall does not remain structurally sound in earthquakes, which could lead to catastrophic failure. Additional earthquake-induced problems potentially exist, because, although the seawall's wood structures are buried in fill, salt water still reaches the wood, and this has allowed gribbles, or marine isopods (think water-dwelling potato bugs about half the size of a grain of rice), as well as shipworms to damage the seawall. The process occurs quite slowly, because the seawall's concrete and steel outer layers make it much harder for the animals to reach the wood, which translates to fewer attackers. In addition, gribble destruction is normally enhanced by waves, which eat away at the weakened wood; thus, wood not battered by waves lasts longer. A weakened seawall is also more susceptible to the rising sea level, another modern threat that didn't concern our predecessors.

I understand that a seawall was viewed as a necessity, but I think we lost a valuable piece of the Seattle story when we built it. When Railroad Avenue rested on its trestles above Elliott Bay, no one doubted they were on a platform that extended out over the water from the city's original shoreline. The water underneath the planks, and the possibility of dropping through a hole to the sea below, were clear evidence that we were manipulating the environment and pushing the shoreline west. The trains and automobiles traveling over the same planks further made it clear that we had done so to suit our economic and transportation needs.

The waters of Elliott Bay were now stopped at the seawall. You couldn't fall through the concrete covering the hundreds of thousands of cubic yards of earth behind the cement. You couldn't tell that the shoreline had once been hundreds of feet east of the concrete seawall. With the new seawall, all vestiges of the historic boundary between land and water had been wiped away. The Seattle shoreline, which had once been a dynamic place of beach, bluff, and tide, was now a line of concrete and wood, where you couldn't even reach the water.

◆ ◆ ◆

Before all of these changes on Railroad Avenue could take place, however, a man arrived in Seattle who would do as much to change the city's landscape as anyone. Eugene Semple was a former lawyer, past editor of Portland's *Oregon Daily Herald*, and the penultimate governor of Washington Territory (1888–89). He had also tried to become Washington's first governor when it became a state on November 11, 1889. Semple's biographer described him as a "typical frontier speculator" with interests in real estate, logging, and mining. "This wiry bespectacled man possessed an entrepreneur's instinct about the region's potential for economic development and most of his life was devoted to cashing in on the potential."[47]

After losing the gubernatorial election—to another former territorial governor—Semple moved to Seattle. He could see that it was becoming the premier city of the Pacific Northwest and an ideal location in which to further his entrepreneurial pursuits. In contrast to that of Thomas Burke, Semple's focus on Seattle was primarily about making money for himself, though he wasn't above a bit of chicanery with allusions to the Seattle Spirit. If the city benefited, that was okay, too, but it was rarely his priority.

His most ambitious moneymaking plan came out of his work as a member of the Washington State Harbor Line Commission. The five-person commission had been established to determine ownership of the state's tidal property. Before attaining statehood, Washington Territory had not actually owned any of its tideflats. Instead, the federal government, as it did with all other territories, had held all submerged lands beneath the ordinary high-tide line in trust for the citizens of the future state.[48] (One effect of the federal jurisdiction is that all the ordinances granting tideflats to the railroads were illegal, which meant that neither the Seattle and Walla Walla, nor the Seattle Lake Shore and Eastern, nor the Northern Pacific had had any legal right to what they had built along the waterfront. Nor did any of the squatters and jumpers, the oyster farmers, or those with federal scrip.)[49]

In October 1890, the commission released its decision, giving the city of Seattle, as opposed to those such as the railroads who already had a presence on the tideflats, broad public control over its own waterfront. Not surprisingly, Thomas Burke was apoplectic. It was an outrage, a monstrous outrage, he told the *P-I*, that the commission would seize the valuable waterfront property, taking it out of the hands of legitimate businesses who had already worked to improve Seattle, such as the ones he had financial connections to, and place it under public control. In the end, though, Burke would have his way, as a gubernatorial change led to a new commission that put the waterfront back in the hands of the railroads. Put simply, the tideflats were on the open market, available to anyone willing to pony up the cash.

One of the advantages of working on the Harbor Line Commission was that it gave Semple a thorough understanding of the changes that came with statehood. He realized that the tideflats constituted an area open to prospective development, particularly if you had friends in the state legislature who would help pass laws that benefited people such as yourself. In 1891, he proposed to fill in and reclaim the entire mouth of the Duwamish River from Beacon Hill to West Seattle. Semple didn't exactly couch his plan as one that would enable him to make money. Instead, he framed the reclamation in terms of the Seattle Spirit. He proposed to complete the long-desired goal of uniting salt water and freshwater via a canal; but instead of the expected route through Lake Union, Semple proposed to cut a canal through Beacon Hill to Lake Washington. (Before Semple's plan, all other proposed canal routes had focused on some sort of connection via Lake Union.) In order to provide the funds needed to excavate a canal through the more than three-hundred-foot-tall hill, Semple would fill in the tideflats and sell liens on the new land.

Formal tideflat-reclamation was the logical next step in developing Seattle. Before Semple's proposed project, filling had been haphazard, opportunistic, and inconsistent. It had generally been done by businesses or individuals—like Dutch Ned, with his wheelbarrow—when they had a chance or a need. Fill deposited in this manner often settled, rotted, and stank. It was also, as seen in the case of the 619 Western building, shot through with air pockets, large and small, making it not the most suitable building surface. Organized, planned, and funded, Semple's method would lead to tightly packed, homogenous fill, not ideal as we now know, but far better than what had been dumped previously.

Seattleites had been making land the old way, not because they didn't understand the importance of using better fill, but because they hadn't any reason to change. What Semple proposed was possible only because of changes in the law following statehood. People and corporations could finally officially own property on the tideflats. And because of legislation pushed through by Semple's friends in the legislature, which opened the door to such development, there was a way to make and sell the new land.

Like so many of his fellow Seattleites, Semple knew that regulated filling of the tideflats was a way to turn Seattle into an "ideal commercial city."[50] It would be enticing to industry because the fill would provide a new, flat landscape to develop, away from the hard-to-build-on hills of downtown. The made land would have deepwater access and railway frontage on the same elevation, and raising buildings on it would be easier and cheaper than on pilings.

This was the city that Seattleites had dreamed of, a place where industry could grow, produce goods, and have access to transportation from land and sea. It would make Seattle a wonder of the world. If we had any doubt that we needed ample flat land, we had to look no further, wrote Semple, than San Francisco, which had faced and overcome similar natural obstructions by doing what he proposed.

Semple revealed his plans to the public through the newly established Seattle and Lake Washington Waterway Company. Not only would they slice a canal through Beacon Hill, Semple's company would also excavate two waterways into the tideflats. The East and West Waterways would run north to south for more than a mile and be a thousand feet wide and twenty-eight feet deep at low tide. To reach the cut through Beacon, the Canal Waterway would run for a mile at a right angle from the East Waterway. The 28 to 30 million cubic yards of soil and mud excavated from the waterways and canal would be used to fill and reclaim the tideflats.[51]

◆ ◆ ◆

In order to get a feel for Semple's great project, I descend from Beacon Hill to another obscure spot deep in the industrial core. Officially known as Jack Perry Memorial Park, the 1.1-acre Port of Seattle property sits at the dead end of a side road running west from Alaskan Way. The park appears to have been carved out of private space. Fences with barbed wire surround two sides of the park. At the end of the road, yellow cement blocks prevent you from continuing west, which is a good idea considering that going farther would plunge you into the East Waterway next to Harbor Island.

If you do walk past the blocks, though, you have a close-up view of one of the central features I have been describing throughout this chapter. Jutting out from the concrete shoreline is a wood pier built atop a wood trestle. Certainly not as grand as the extensive trestles that made up Railroad Avenue or even Joe Surber's relatively small trestle across the tideflats, this one at least gives you an idea of what they looked like: twelve-inch-wide piles, covered up to the high-tide line in barnacles, mussels, and fingerlike seaweed; ten-inch-by-ten-inch pile caps (the beam running across the top of the piles) topped at a right angle by same-sized stringers; and four-by-ten-inch planking. If this structure had supported a railroad, the rails would have sat on wooden ties that spanned the stringers.

I have come to this little park because of what happened near here on July 29, 1895, when dozens of dignitaries stood on a smoke-billowing dredge and watched what the *Seattle Times* crowned as "undoubtedly the most important work ever undertaken on Puget Sound."[52] Or at least the greatest since that other remarkable day, the May Day picnic of 1874, when those naively optimistic strivers had set Seattle on a path to having transcontinental rail (which arrived in 1893), to becoming the West Coast's largest coal supplier, and to amassing a population of fifty-five thousand. One other significant difference between this day and that earlier one: in 1874, Seattleites had come to work; in 1895, they came to watch; but everyone still had the same spirit of gathering together to see a new Seattle grow.

They stood by the hundreds on a nearby scow anchored to the dredge, with thousands more watching from docks.[53] Everyone who wasn't on a dock had had to travel by boat to this spot, because a high tide covered the tideflats and the boats were floating in twenty feet of water. The nearest spot of land to the west, West Seattle, was a mile and a half away. To the east, about a half mile distant, lay the docks and wharves that reached south into the tideflats; and two miles to the south was the mouth of the Duwamish River.

Aptly named the *Anaconda*, the dredge was 110 feet long and 32 feet wide. Its most prominent features were the tall smokestack and curi-

FIG. 3.8. FILLING THE TIDEFLATS, JULY 29, 1895. Mud rushing out the pipe of the dredger *Anaconda* marks the start of the organized filling of the Duwamish River tideflats. Behind the spectators stand a shipyard, sawmill, bottling works, and furniture factory. Today, Alaskan Way occupies this location, and CenturyLink Field stands behind it. Courtesy University of Washington Special Collections, UW 5461.

ous front end, which looked like a battering ram or the bow of a Viking ship. It was actually a hoist holding a hollow cutting tool consisting of a seven-blade spinning cutter inside a thirteen-blade rotating cutter. Extending out of the hollow cutter was a suction pipe that connected to a huge centrifugal pump. Built to suck up sediment, the *Anaconda* could discharge ten thousand cubic yards of material over a twenty-four-hour period.

Standing in the upper-deck engine room of the dredge was the day's master of ceremonies, Eugene Semple, and his twenty-two-year-old daughter, Zoe. At 11:29 A.M., Miss Semple stepped onto a wooden block, reached toward the ceiling, grabbed a small lever with her white-suede-gloved hand, and pulled. As a deafening wave of boat whistles blasted across the bay, the dredge began to shake and sway as great cogs turned and set in motion the blades of the cutting tool that were sunk into the mud twenty-five feet below the surface of Elliott Bay. Within seconds, clear water, then black mud, exploded from joints of the suction pipe. The explosion quickly dissipated as the underwater jaws continued to eat away at the Duwamish mud and the pump began to suck the water out to a stationary pipe suspended above the tideflats. The twenty-seven-hundred-foot pipe made a graceful arc curving east and northeast on triangle-shaped trestles to a section of tideflats just west of the city's southernmost wharves and piers. In modern terms, this would be a block or so west of Safeco Field.

A photograph taken at the time shows onlookers crowding every available spot on the piers surrounding the discharge pipe. Also visible are one man and two boys who climbed out to the end of the pipe, about six feet above the water. One of the youngsters stuck his hands in the black stream but drew them back quickly, displaying a fish he caught. Its journey through the pipe was "taken by many as most significant."[54]

If the pipe-climbing lad had not caught it, the fish would have ended up in an area recently protected by a brush bulkhead, made of pilings, brush bundles, and trees. The pipe would continue to supply sand and silt until it built up the fill to a height two feet above high tide. After the new land had been created, the pipe would be moved to the next location. Brush was used instead of a more substantial material because it allowed water to escape—though not all the water escaped, which led to problems we still experience.

Semple's plans called for building nine miles of bulkhead, requiring an estimated sixteen thousand piles, forty thousand tie-back poles, and 180,000 cords of brush. It would provide many jobs for those who needed them following the Panic of 1893, including jobs for thirty-five to forty men working the dredges, thirty to thirty-five cutting brush, and another twenty-five building the bulkhead. By June 1897, the *Anaconda* and its sister dredge, the *Python*, had excavated three thousand feet of the East Waterway to a depth of thirty-five feet below low water. The 1.7

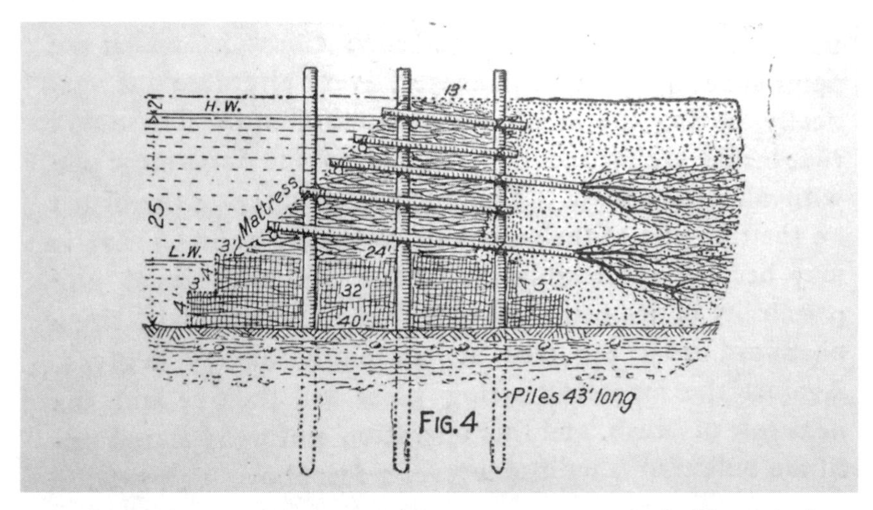

FIG. 3.9. BULKHEADS FOR FILLING IN TIDEFLATS, 1895. The bulkheads built to contain the dredged-up mud and sand from the Duwamish River were carefully engineered structures. The word *Mattress* refers to the sticks laid directly on the mudflats. *H.W.* refers to high water and *L.W.* to low water. From *Engineering Record* 32, no. 11 (October 19, 1895): 363.

million cubic yards of sucked-up muck had created seventy-five acres of new land.

Dredging did not proceed, however, without the requisite legal challenges and subsequent delays, or what we might now call the Seattle Process: the grand vision took two steps forward, and legal manipulation pushed it three steps back. After nearly two years of work cessation caused by legal and financial challenges, dredging resumed in October 1900. By September 1904, Semple's company had finished the East Waterway and begun work on both the West Waterway and Canal Waterway. The 8 million cubic yards of excavated material created 333.26 acres of new land.

◆ ◆ ◆

One of the best ways to understand the effects of Semple's dredging operation is to take a look at the block or two due west of CenturyLink Field. Known as the Dearborn South Tideland Site, it was excavated by archaeologists during the initial digging of the entry point to the State Route

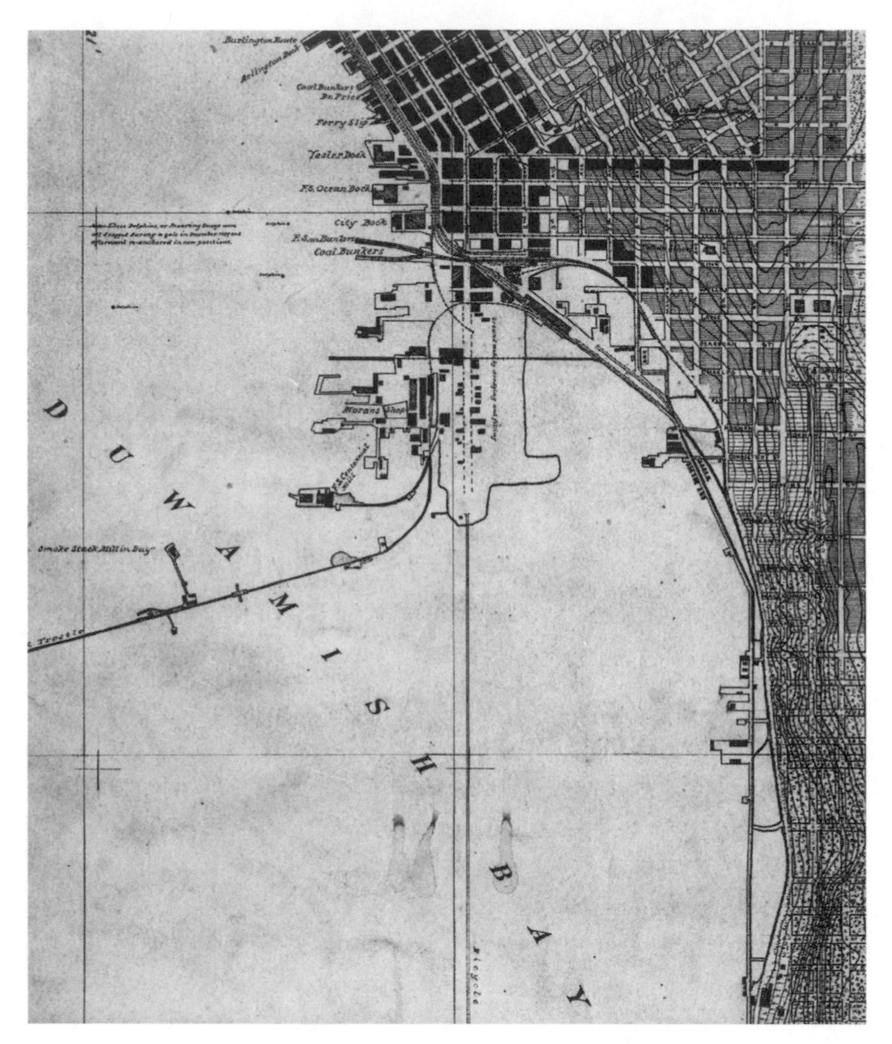

FIG. 3.10. DETAIL OF THE U.S. COAST AND GEODETIC SURVEY OF SEATTLE BAY AND CITY, 1899. At tideflats newly filled with dirt from Eugene Semple's dredging project, Moran's shipyard and the Centennial Mill were the first large buildings to rise on the new land. Note how the train tracks cut across the tideflats and run along its margin, as well as extend north along Elliott Bay. The line running due south of the made land is a bike path, built atop an old trestle that runs over what is now First Avenue South. Library of Congress, Geography and Map Division.

99 tunnel. In 2009, a team focused on two trenches that ran the entire length of the site, from Royal Brougham Way to South Dearborn Street. The trenches were dug to hold transmission lines formerly attached to the Alaskan Way Viaduct. A more detailed study of additional locations within the study site followed a year later.

At the base of the trenches, archaeologists found gray silt and sand interspersed with shell fragments. Typically reached during excavation around six feet below the surface, this was sand and silt chewed out of the East Waterway by the *Anaconda* and *Python*. The archaeologists did not dig below this level, although they did find truncated pilings that extended up into the fill, a clear indication of the network of wood posts that supported the piers and wharves built before 1895.

As Semple had claimed would happen, the new homogenous fill from the Duwamish dramatically changed how people built on the tideflats. This can be seen at either end of the project site. At the north end, archaeologists found a twenty-inch-thick lumber foundation buried nine feet below the surface, in a trench dug four and a half feet deep into fill. The foundation supported a five-foot-high vertical brick wall, which had once been part of the United Warehouse Company building, a two-story, 275-foot-long warehouse erected soon after the fill had been deposited. At the south end of the site were additional brick walls and concrete foundations. These were part of the slightly larger American Steel and Wire Company warehouse and office. In the middle of the site, dig teams unearthed the concrete-and-brick remains of the Washington Saw Works and Seattle Ice Company buildings.

Industry leaders not only found the fill cheaper and easier than pilings to build on, but they also recognized that they could build more substantial buildings. Before the fill, most buildings on the piers were made of wood; brick and concrete would have been less stable on the pilings, particularly given the pilings' susceptibility to shipworms.[55] In the post–Great Fire world, when people understood the downside of building with wood, brick buildings carried the advantage of being less susceptible to fire than wooden ones.

Archaeologists have uncovered remains that tell two additional stories. The first comes from one of the trenches cut for the new transmission lines. Seven feet below the modern surface lay a scattering of items, including white ceramic dishware, glass tumblers, and cut-up mammal bones. They were situated along the back side of what had been the First

Avenue Hotel, near the kitchen, which served a dining room and saloon. Other found objects in the dig area included three items that wouldn't look out of place in modern Seattle: a Rainier Beer bottle, a bottle of mineral water from springs on Zarembo Island in southeast Alaska, and a bottle of Mellin's Food for Infants.

What these items help illustrate is that the spread of industry onto this made land did not limit who would take advantage of it. People still needed cheap places to live. They found rooms in hotels, tenements, and boardinghouses, ate in restaurants, and frequented saloons and a beer garden. Although the 1900 census data shows that the majority of people who resided on the tidelands were single men, there were a few families, as implied by the baby formula.

The census data also shows a change in who lived on the marginal lands in Seattle. In contrast to the 1880s and earlier, when Chinese and Native residents dominated, the community of 1900 was far more cosmopolitan. The census taker found people from at least twenty-five countries—primarily Europeans but also Syrian peddlers, a New Zealand ship carpenter, and a Peruvian sailor—some of whom, along with scores of Finnish miners, had probably come to Seattle in search of Klondike gold. The majority, however, were tradesmen doing everything from making cigars to running a printing press to molding iron.

Not everyone appreciated those who dwelt in what the census listed as "houses, boats, houseboats, shacks, tents, barges, etc. on the water front." The shacks' location on the approaches to the city meant that "visitors to the city . . . received a bad impression."[56] In May 1907, after several years of attempts by improvement clubs and city officials to "beautify the city," the city board of health executed the "biggest cleanup in the history of Seattle," when inspectors spread across the tideflats condemning the unsightly shacks. By June the crusade had "borne excellent fruit," with more than 120 shacks having been eliminated.

This fait accompli is the second story revealed in the dig. Before the archaeologists could reach Semple's gray, well-packed fill, they had to dig down through a foot or two of gravel and sand topped by asphalt, followed by several feet of what the archaeologists called demolition debris or industrial fill. It contained a mixture of dirt, wood scraps, stucco, slag, bricks, charcoal, and coal, sometimes in layers and sometimes in pockets, but it was not purely the remains of a bygone industry. Archaeologists also uncovered domestic and personal items mixed within the industrial materials, looking as if perhaps the household goods had been abandoned by tenants when they left the buildings for the final time.

This abandonment had taken place in early 1908, when the Union Pacific Railroad began to raze all buildings on the site and deposit the material as fill, which was necessary in order to raise the ground surface up to the level of adjacent railroad tracks. By January 1910, all of the earlier buildings were gone, replaced by rows of train tracks and a warehouse, a sight that many considered far better than rows of unsightly shacks.

The astounding transformation of the three blocks of the Dearborn South Tideland Site is a bit unusual in its speed and scope, but it does highlight the development of the tideflats and, in particular, the importance of railroads to the city and the tidelands. For example, if, in 1912, you had attempted to walk east along or a bit north of Dearborn Street, you would have crossed at least fifty sets of tracks by the time you reached Fifth Avenue South. Along the way, you would have stepped over the tracks of the Northern Pacific and Great Northern, which fed into James Hill's King Street Station and north to his tunnel under Seattle, and over the tracks of the Union Pacific and Seattle's fourth transcontinental train, the Chicago, Milwaukee, Saint Paul, and Pacific Railroad. Their tracks ended at the terminal they shared, a block east of the King Street Station. To do the same walk today, you would cross five sets of tracks; only three train companies now operate in Seattle, and their main yards are on less-desired land south and north of the central business district.

◆ ◆ ◆

Filling in the tideflats was only half of Semple's grand scheme. He still had to cut his three-hundred-foot-deep canal through Beacon Hill. The South Canal, as it was called, was the most impractical and controversial part of his plans. His route from the tideflats to Lake Washington was the shortest connection between salt water and freshwater, but did he really think it was possible to slice a canyon through the hill? I have never seen any records indicating that he knew the geology of Beacon Hill and what it would take to open it up. But he did know that the millions of cubic yards of material could be used to fill in the tideflats, which he could then sell to industry and other speculators. Slicing through Beacon Hill may seem in modern terms to be completely foolish, unrealistic,

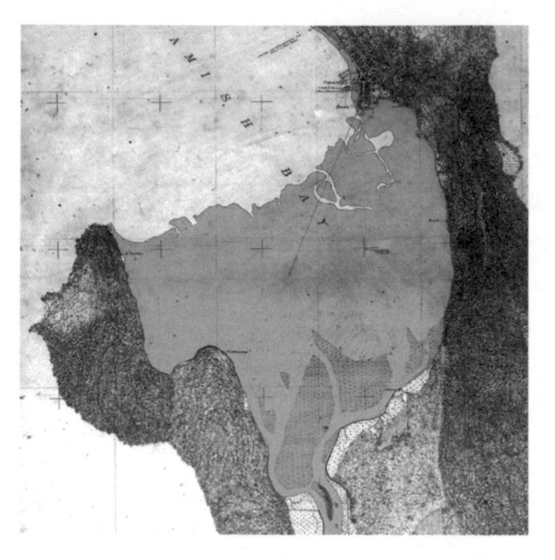

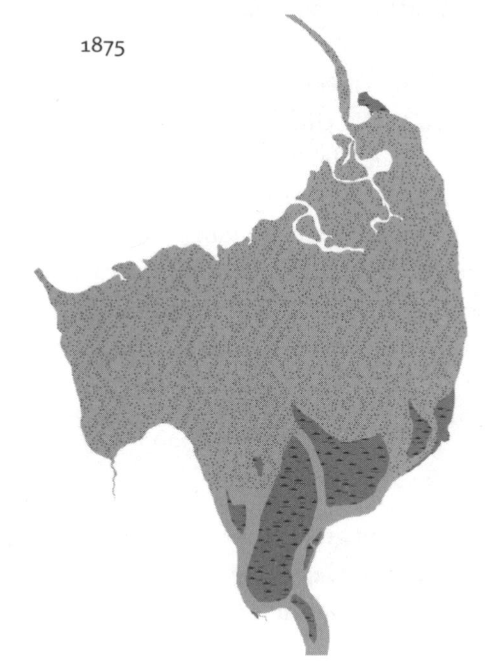

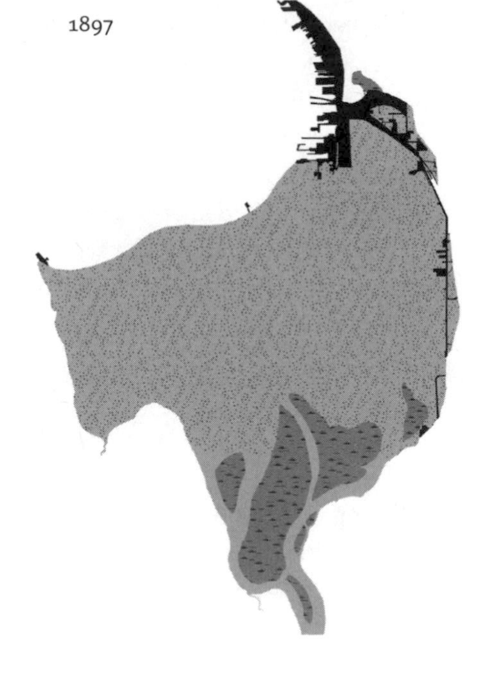

FIG. 3.11. EVOLUTION OF THE TIDEFLATS: 1875 TO THE PRESENT. Adapted from the Waterlines Project.

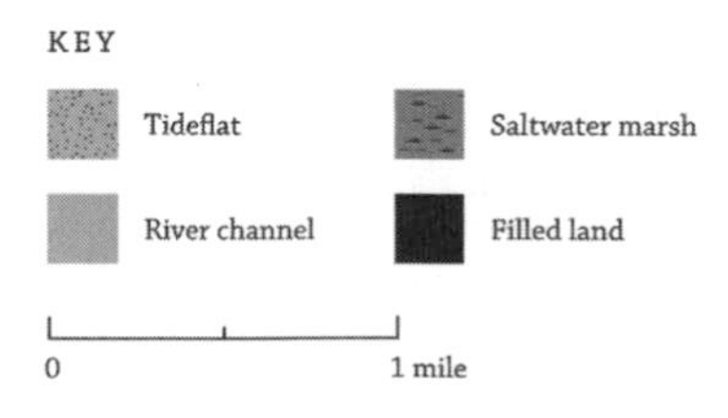

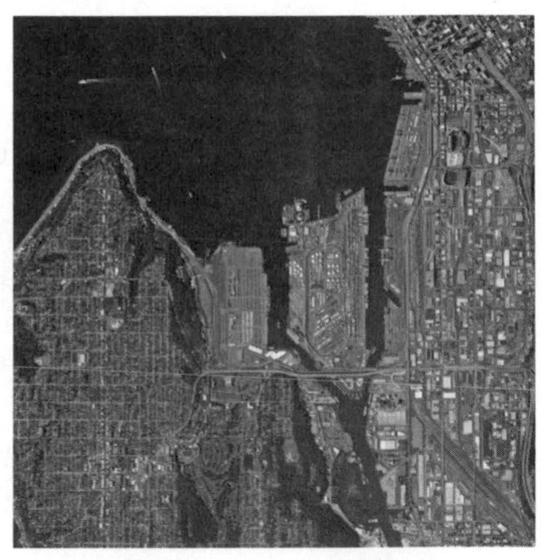

1909

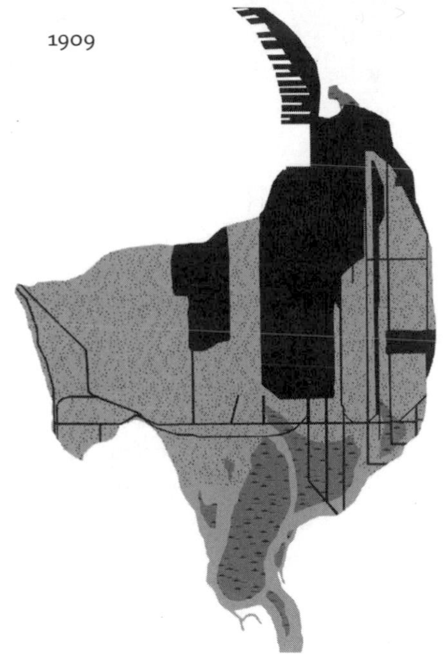

Present

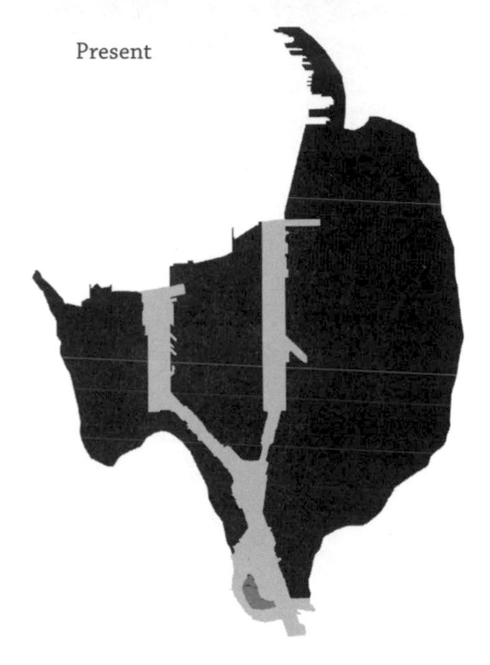

and environmentally asinine, but in our predecessors' eyes it was standard operating procedure and a sign of progress.

Semple's crews began cutting the South Canal on November 15, 1901, at the base of Beacon Hill, directly east of the Bayview Brewery at Tenth Avenue South and South Hanford Street.[57] Semple's plan was to draw water from the Beacon Hill reservoir and send it through wood-stave pipes into a high-pressure steel pipe that dropped nearly three hundred feet to the base of the hill. As the water plunged downslope, the pipe narrowed, increasing the water pressure to the point that it would "be shot into the hillside at a cannon ball rate."[58] During a typical day, the water gun blasted Beacon Hill with 14 million gallons of water and generated about two thousand cubic yards of fill. More than two miles of flumes then distributed what had been Beacon Hill to what would become made land. By late 1904, Semple's workers had washed away seven blocks of Beacon Hill and created fifty-two acres of solid land along the hill's western edge.

Although the value of made land in the former tideflats had increased at least tenfold since Semple's project began, not everyone supported the South Canal. Opponents claimed that Semple actually had no plan to build his canal; instead, he was misleading the public and only blasting away at Beacon Hill to create material to fill the tideflats, which he intended to sell. Ironically, those who opposed the canal supported filling in the tideflats; what they didn't approve of was that Semple used a public project to benefit himself. In addition, many of Semple's opponents, including Burke, supported and owned property near where a north canal could be built (where the modern ship canal now exists).

Yielding to public opinion and the north-canal plan's powerful supporters, the city council voted to turn off the supply of water that Semple needed to run his hydraulic cannons. By the end of 1904, notes Semple's biographer, weeds were growing in the chasm on Beacon Hill. Semple's decision to try to cut through Beacon Hill ultimately had been his downfall. In May 1905, he was forced to resign as president of the Seattle and Lake Washington Waterway Company, though the company continued to exist and, by 1917, had filled in 92 percent of the tideflats. No more work would be done ever again on the proposed South Canal through Beacon Hill, but if you look, you can still find two infrastructure elements from Semple's misguided scheme.

To see the first, go to the spider's web of ramps and overpasses that connect the Spokane Street Viaduct and Interstate 5 to Columbia Way and Beacon Hill. Engineers chose this spot to build this interchange because it is where Semple had started his canal, creating a large, unoccupied gap that eventually provided the easiest access up the hill. The second is down on the flats. In contrast to the typical blocks measuring more than seven hundred feet long, South Hinds and South Horton Streets are just three hundred feet apart. They are so anomalously close because they mark the north and south boundaries of what was to be Semple's Canal Waterway, which would have run from the East Waterway to his canal through Beacon Hill.[59]

More significant than these two spots is a long-term problem created by Semple's work on filling in the tideflats. Fortunately, the problem is not something we encounter on a regular basis. The first time that Seattleites experienced it was just before noon on April 13, 1949, when a magnitude 7.1 earthquake hit the Puget Lowland. At what is now the Starbucks headquarters and was then a Sears, Roebuck building, geysers of mud and water erupted as high as three feet. Flowing for as long as twenty-four hours, the sand slurries filled basements around Sears.

The same phenomenon happened again, on February 28, 2001, during the Nisqually quake. Most of the sand geysers were discovered only after the event, but Bob Norris, a seismologist formerly with the U.S. Geological Survey, was lucky enough to witness one of the mini volcanoes of sand and water. He heard it first, while standing on Harbor Island. "It sounded like running water or something squishy. Rising out of the ground was a mound of brown water. It soon broke into a geyser about a yard high, which would get clogged, blow out, and clog and blow out again."[60] Within a short time, the slurry had become a pool about twenty feet across. When he returned several days later, he found a forty-five-foot-wide conical mound of dark sand.

University of Washington geologist Kathy Troost, who spent the days after the Nisqually quake surveying these sand boils, told me that she found hundreds of the tiny eruptions, some of which left behind splatter marks higher than her head. One hard-hit area was the north end of Boeing Field, which was built atop several old meanders of the Duwamish River. Troost found bricks, chunks of coal, blocks of wood, and a twenty-one-pound clump of fused slag and bricks, all of which had been used as fill and which had jiggled up and out during the earthquake.

Known generically as liquefaction, and not limited to earthquakes, such bursts and eruptions form during the shaking of water-saturated sediment. As the pressure builds, the sand begins to lose its strength and turns soupy, similar to when you wiggle your toes in wet sand and your foot begins to sink in what looks like a pool of dirty water. When the pressure is too confined, one method of relief is for the water to burst out of the ground. Liquefaction can destroy buildings when foundations sink into the weakened ground, which happened to many structures on the built land of San Francisco's Mission and Marina Districts, and at the Oakland airport, during the 1989 Loma Prieta earthquake.

No major buildings have collapsed in Seattle's earthquakes, but in those events the most significant damage to the city's buildings occurred on the made lands of the tideflats, where bricks fell, sidewalks and streets cracked, chimneys collapsed, and bridges got stuck in either an open or a closed position. Many streets also flooded from water lines breaking and from water ejected during liquefaction. (In the 1906 San Francisco earthquake, liquefaction-ruptured water pipes prevented firefighters from obtaining water to stop the spread of the blaze.) In addition, the shaking severed electrical lines buried in the disturbed fill. This is one of the central consequences of our wholesale alteration of the tideflats: the made land is unsafe during earthquakes. The issue is further exacerbated because fill can prolong the shaking from a quake, making buildings more susceptible to failure and potentially harder to reach by emergency personnel because of the flooded roads and widespread sand boils. Seattle has improved its seismic building codes and spent millions retrofitting buildings, but when I asked geologists where they would least like to be during an earthquake, all said the former tideflats.

Despite the potential problems of liquefaction, I know that Seattle could not have become the city it did without eliminating its tideflats. Try to imagine the city without the fill. How would we access downtown from the south? Traffic is challenging enough now. I don't want to consider how bad it would be without the multiple routes afforded by the extensive roads on the fill. Or what about manufacturing? Although we often overlook the city's industrial base, it provides significant economic benefits and could not easily have been developed anywhere except on the fill. And what sort of city would we be without two massive sports arenas sitting on fill?

Looking back at our history, it was inevitable that we would change the tideflats. As Seattle grew from a handful of settlers dumping sawdust to a town where the entire population would gather for a quixotic dream to a city ambitious to become world-class, every generation believed that its path to prosperity lay through the expanse of mud and sea that hemmed it in. They were correct. Of all the large-scale landscape-changing projects I discuss, filling in the tideflats is the only one that was topographically necessary.

There's a problem, though. The tideflats were shaped by thousands of years of earthquakes, volcanic eruptions, and mudflows, as well as by the daily drama of the tides. It was landscape ruled by the vagaries of geology, where nothing was static and change was constant. I worry that people such as Semple and the May Day picnickers of 1874 did such a good job of burying these stories under garbage and sawdust and sand and asphalt and concrete and buildings that they have blinded many Seattleites to the instability of our former topography. Fortunately, there are those who have begun to understand the consequences of the tideflats' former dynamic nature and are beginning to point us in a safer direction. I can only hope their message is heeded. But we can never forget that the land does not forget, and that with all of this unstable fill, the consequences could be catastrophic.

Replumbing the Lakes

THE EARLIEST VISION FOR URBAN PLANNING AND LANDSCAPE change in Seattle occurred at a Fourth of July speech in 1854. Speaking on or near the property he had recently homesteaded, Thomas Mercer proposed bestowing the name of Washington on what was then known as Lake Geneva, Lake Duwamish, and *hyas chuck*, or "big waters." He then made known his idea for a future Seattle, suggesting the name Union for the smaller lake, where everyone had gathered, noting the possibility "of this little body of water sometime providing a connecting link uniting the larger lake and Puget Sound."[1]

What Mercer prophesied was a joining together of the two great bodies of water that gave Seattle its hourglass shape. Between Lake Washington and Puget Sound lay Lake Union, which connected to the sound through Salmon Bay via Ross Creek, but which was separated from the larger lake by a half-mile-wide neck of land. Lake Union was about nine feet lower than Lake Washington and twenty-one feet higher than the sound.

We don't know how Mercer envisioned connecting salt water and freshwater; but given the era, some sort of canal-and-lock system is logical. Following upon the success of the Erie Canal, which opened in 1825, planners had proposed and, by the time of Mercer's speech, built more than three thousand miles of canals across the eastern United States. Those waterways had became an indispensable part of American society by improving transportation and facilitating industrialization, two goals never far from the minds of Seattleites. Canals were also emblematic of the nineteenth-century belief in technology, or what the

Union Bay, circa 1916 (detail)

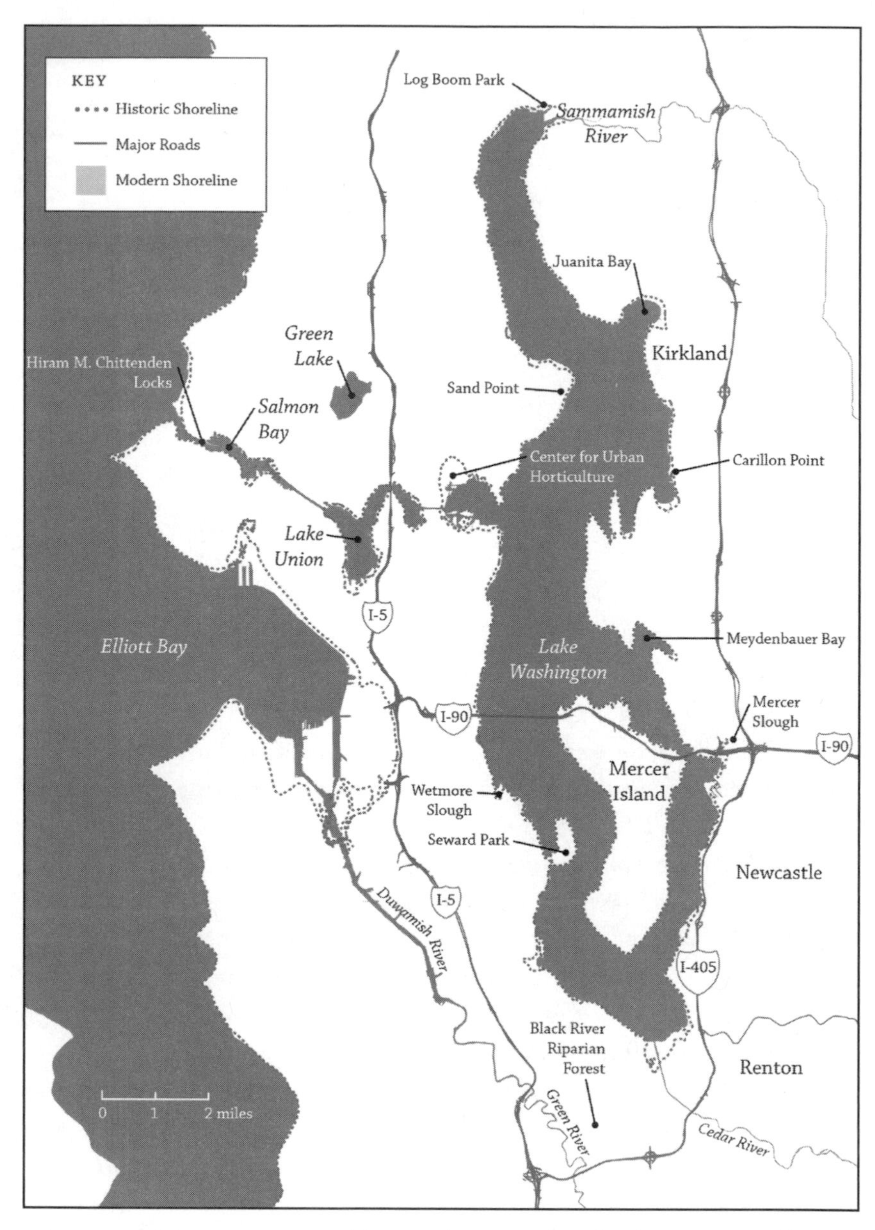

MAP 4.1. LAKE WASHINGTON. Historical mapping by UW's Puget Sound River History Project.

historian Matt Klingle calls "a fetish, endowing it with miraculous powers," in his well-regarded *Emerald City: An Environmental History of Seattle*.[2] Canals and railroads, in particular, had changed the country and the countryside, transforming the physical environment and delivering endless material abundance.

With their canal, Seattleites hoped to make the city into an ideal port for trade. The integrated water bodies would have a direct link to the Pacific Ocean and would supply the city's main trading partners with a hospitable environment: ships would be untroubled by the dreaded shipworms, which couldn't survive in freshwater. Advocates further hoped that opening up the lakes would lead to industrial development, since freshwater was less corrosive and the lack of tides made it easier to load and unload vessels. With more than eighty miles of shoreline, Lake Washington and Lake Union would also have enough room for a freshwater port for the U.S. Navy, which had long sought such a facility on the West Coast.

Mercer's idea may have been one that early Seattleites could cotton to, but it was not one they could actually act upon, although they tried. The first to try was Harvey L. Pike, who, wielding little more than his ambition and some garden implements, made a futile attempt in 1860. Recognizing the potential for a navy port, the federal government decided to investigate unification of the two lakes a decade later. The region offered one of only three potential places on the Pacific Coast to build a secure port for the navy, but a report concluded that Seattle had too few people—about fifteen hundred—and insufficient resources to justify further study. In 1882, landowners around Lake Union and Salmon Bay, including Thomas Burke, decided once again that Seattleites could do it alone without outside money, and they hired a Chinese labor contractor to link the lakes. The workmen began by widening and straightening Ross Creek and then dug a sixteen-foot-wide ditch at the Portage, where State Route 520 now runs. Both cuts required small locks to control water as it dropped from Lake Washington down to sea level.

In spite of this initial success, local canal-backers discovered that they actually needed the federal government, or at least its money, to build canals large enough for more than just logs. In 1892, Congress published a report that analyzed five different routes, but it would not provide money to start the project. The one route the government didn't suggest was Eugene Semple's grand plan to cut a canal through

Beacon Hill, which perhaps illustrates how misguided Semple was. Despite the influx of money and people following the Klondike gold rush (the 1903 population of 115,000 was about double the population of 1897), another federal report, in 1903, concluded that Seattle was still not ready for a canal.

Finally, in 1906, a man arrived who would quickly and permanently leave his mark on Seattle and its long-envisioned plan to connect salt water and freshwater. Hiram M. Chittenden had spent most of his career with the Army Corps of Engineers before his assignment to Seattle. His biographer describes Chittenden as "a dedicated, highly intelligent, inhumanely industrious man . . . [who] fixed immediately upon the Lake Washington canal as the most important project in his district."[3]

By 1910, Chittenden had convinced Congress to authorize $2,275,000 for construction of the locks. Just seven years had passed since the most recent federal rejection of Seattle's canal plans, but the city had changed significantly. Its population had more than doubled, to 237,000 people. It had also nearly doubled physically with the annexations of Ballard, Georgetown, West Seattle, Ravenna, Laurelhurst, and most of southeast Seattle. For the first time ever, the changes entailed in building the ship canal would actually be within Seattle city limits.

Construction began on the two sets of locks and the two sections of canal (Lake Washington to Lake Union and Lake Union to Salmon Bay) in autumn 1911. The locks' gates closed on July 12, 1916, and thirteen days later Salmon Bay had risen to its present level of twenty-one feet above mean sea level. In late August, the barrier separating Lake Washington and Lake Union was breached. Water drained for four months, until the water levels equilibrated through the two canal sections, which together were a little over a mile and a half long.

Officially known as the Hiram M. Chittenden Locks, the structure consists of a spillway dam, which keeps the water level in Salmon Bay between 20 and 22 feet above sea level; a fish ladder, which the Army Corps of Engineers renovated in 1976; and two parallel locks: the larger is half as wide as a football field and more than twice as long, and the smaller lock measures 30 by 150 feet. At each end of the locks are two wooden gates. When a set of gates is open, boats enter or leave the lock. When the gates are closed, valves on the sides of the chamber drain or fill it, either to raise boats to the level of Salmon Bay or to drop them to saltwater level.

More than fifty thousand people attended the official grand-opening dedication, carnival, and celebration on July 4, 1917. Never ones to miss an opportunity, the local papers get high marks in hyperbole. In the words of the *Seattle Times*, "Every thinking person on the whole canal right-of-way realized in full—that here, completed, ready for use, actually in use, was a thing that will do more toward bringing Seattle its destined million inhabitants and undisputed Pacific Coast supremacy than any other factor the city has ever known or is likely to know in the present generation."[4]

◆ ◆ ◆

Although one may not initially think of building the ship canal as reshaping Seattle topography, the consequences were as dramatic as those of the regrades and the filling in of the tideflats. Lowering Lake Washington by nine feet reduced the surface area by two square miles, shrank the shoreline by more than ten miles, and eliminated two islands. Or put another way, the change led to the addition of more than thirteen hundred acres of newly exposed land around the lake, most of which has since been built upon; primarily residences have been constructed but also businesses, roads, and parks.

In one fundamental way, though, the story of the canal and locks differs from the story of the tideflats and regrades. For these, the history of the endeavors is integral to understanding how and why they took place. For the ship canal, the story becomes most fascinating after the point when the locks were completed, in part because, despite the sixty-four years between conception and completion, the physical building of the canal and locks took less than a decade. In addition, they caused changes far different from those that their early proponents wanted or could have predicted.

Those changes can be sorted into two categories: economic and ecologic. More than 93 percent of Lake Washington's wetlands dried out and disappeared after the lowering of the lake. In addition, the lake's surface level dropped below the level of its outlet at the Black River, which led to the Black drying out and disappearing. Other changes included the transformation of Salmon Bay from a dynamic tidal ecosystem, comparable to parts of the mouth of the Duwamish River, to a bay permanently under more than twenty feet of freshwater, and the forcing of coho, chinook, sockeye, steelhead, and other species to find a new migration route to the lake. The ecological changes were both abrupt and not felt until decades later.

From an economic standpoint, the changes resulted primarily from the new transportation route created by the canal and locks. Before 1916, boats had moved between salt water and freshwater by making the arduous twenty-mile trip from Elliott Bay up the Duwamish and Black Rivers to the lake. For example, from Kenmore at the head of Lake Washington to downtown Seattle was a forty-mile trip. After the locks, it was a twenty-four-mile trip, and one not complicated by rivers whose volume could change radically or whose route was overgrown and subject to tidal flows. As with the ecological changes, the economic changes came in two varieties: immediate and long-lasting.

(Throughout this chapter I refer to a map and a twenty-two-part newspaper series. The map is a 1905 chart of Lake Washington by the U.S. Coast and Geodetic Survey.[5] It shows the topography of the surrounding lands and the bathymetry of the lake, based on surveys done between 1856 and 1904. It is the most accurate pre-1916 map of the lake. The series of articles was written by historian Lucile McDonald and published in the *Seattle Times* between October 2, 1955, and February 26, 1956. Based on extensive research and correspondence with many long-term residents, some of whom lived on the lakeshore in the 1800s, McDonald's articles provide a human touch that complements the map.)

◆ ◆ ◆

The most significant postcanal ecological change took place at the south end of the lake, about nine miles by air from downtown Seattle. Before the canal opened, much of what is now modern Renton, including the massive buildings of the Boeing plant, the acres of surrounding parking lots, Renton Municipal Airport, and Renton Stadium, was once underwater or part of a several-hundred-acre marsh, according to my 1905 map. At the time, Renton had a population of about two thousand, more than four times what it had when the Seattle and Walla Walla Railroad and Transportation Company first reached town in 1877.[6]

The map also shows that the water at the south end was shallow, less than ten feet deep, so when the lake dropped it exposed hundreds of acres of bottomland. After 1916, the area was partially filled and used

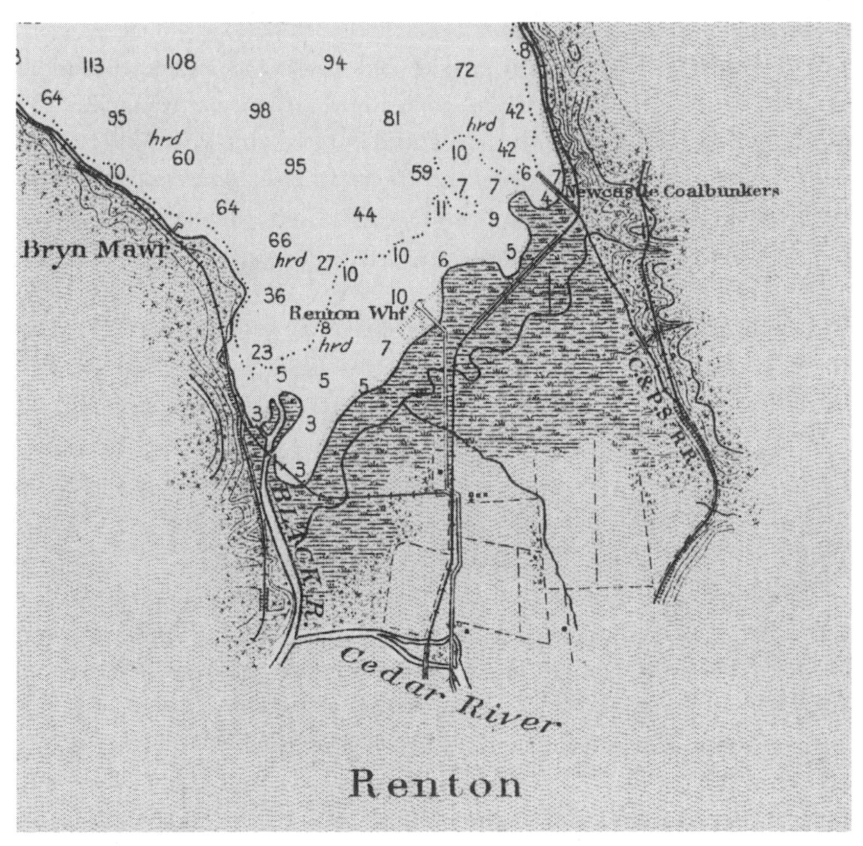

FIG. 4.1. LAKE WASHINGTON TOPOGRAPHICAL MAP, DETAIL OF RENTON, 1904. Courtesy University of Washington Libraries, Map Collection.

as a plant nursery, sawmill, and swimming beach. The big change came during World War II, when Boeing started to expand. To create the new land, builders used material from a five-hundred-foot-long slag heap formed during the peak years of Renton's coal boom at the turn of the nineteenth century. I suspect this means that one might not want to know exactly what is under all the concrete, though it might be better underfoot than aboveground; for years the slag had burned slowly, leaving Renton "overhung by a gray pall."[7]

Cutting through the modern fill is the Cedar River. The river starts about five miles southwest of Snoqualmie Pass and flows through second-growth forest about twenty-five miles to a small dam, where about a quarter of the water gets diverted into pipes to become the main source of drinking water for Seattle. The river then continues another twenty-two miles to the lake. Before 1912, the Cedar River did not drain into Lake Washington. Instead, it turned west, near Renton Stadium, ran for a few hundred feet, and entered the Black River about where the modern Airport Way intersects Rainier Avenue. The Black, the only outlet for the lake, ran parallel to Rainier for a few more blocks before angling west and continuing about three miles to its confluence with the Duwamish River.

According to McDonald, the confluence of the Black and Cedar was the site of the first sawmill in King County outside of Seattle. Started in 1854 by three settlers, the sawmill lasted only two years. Across from the mill was an outcrop of black rock that may have been the earliest-recorded location of coal in the region. The land eventually ended up under the ownership of the Renton Coal Company, named for Captain William Renton, who helped finance the endeavor.

The routes of the Black and Cedar, and the lake's extensive marshes, account for an odd aspect of Renton's geography: the town was not built on its waterfront. Compare the central business districts of Seattle or Kirkland; both sit on the water, whereas historic downtown Renton is more than a mile from Lake Washington. The founders had no other choice. They couldn't build near what should be a better site for industry and transportation—on the edge of the lake, where they could establish piers and wharves—because of the unstable terrain of the marshes.

Throughout the early history of Renton, the Cedar River regularly abandoned its sinuous channel and flowed through homes and businesses. Following a severe flood in November 1909, a group of Renton's civic leaders decided to fix the recalcitrant river by putting it in a straitjacket. Nearly three years of haggling with state legislators followed, but work finally began on June 1, 1912, to straighten and dredge the Cedar and send it directly into the lake. The new seventy-six-hundred-foot channel would remove "forever the menace of floods," promised the *Renton Herald*.[8]

Diverting the Cedar away from the Black was not an original idea. In 1907, Hiram Chittenden had suggested a similar plan, in part to relieve flooding in the Duwamish River valley caused by the Cedar's floodwaters surging down the Black River into the Duwamish. What particularly troubled Chittenden, and those who lived downstream, was that

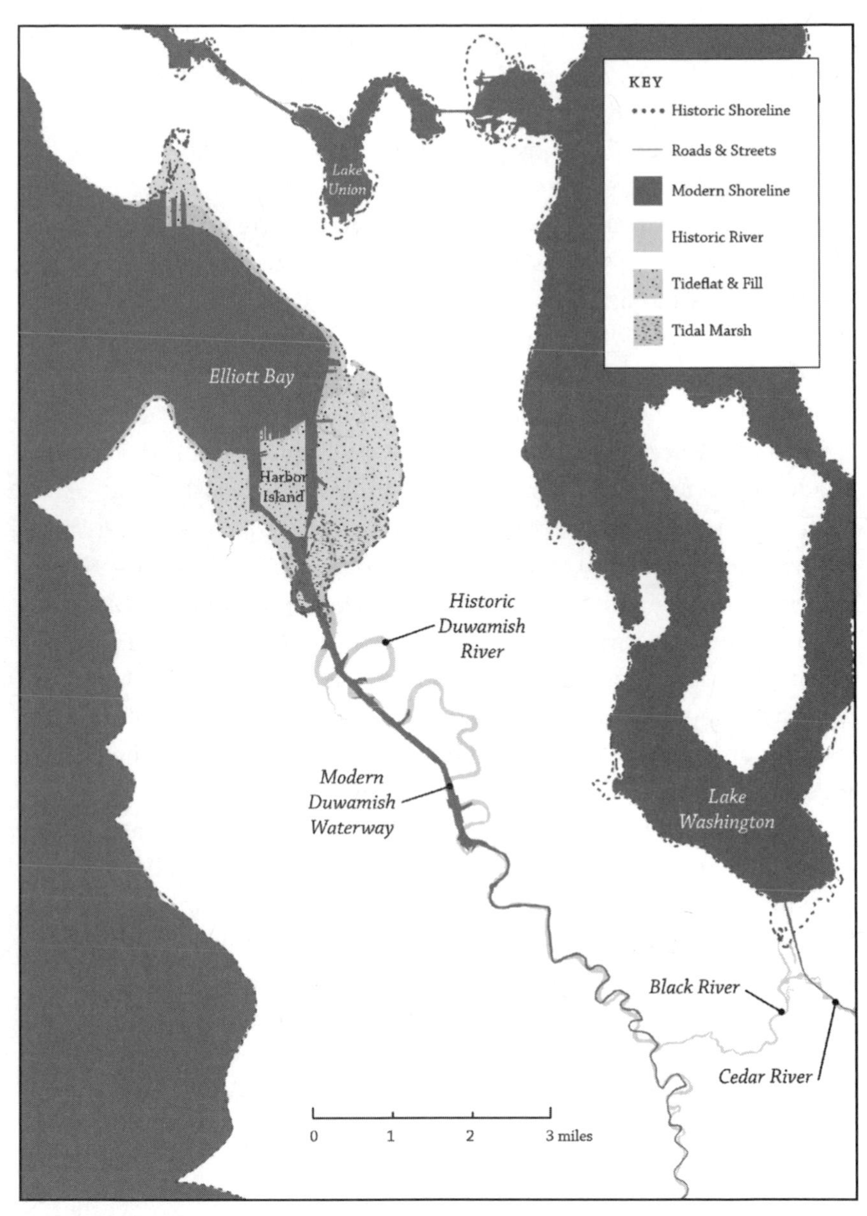

MAP 4.2. DUWAMISH RIVER VALLEY. Historical mapping by UW's Puget Sound River History Project.

after the Cedar had lowered back to its normal level following a flood, the Black would continue flooding for weeks because it drained the lake, which also contained more water than normal from the Cedar and Sammamish Rivers and the many creeks that entered the lake.

Flooding had long been a central concern for people around Lake Washington and in the Duwamish River corridor. As early as the 1860s, and continuing regularly for decades, residents had prevailed upon their elected officials to fix this scourge caused by excess runoff. Reengineering the Cedar would eliminate half of this problem; building the ship canal, which had long been a central goal of those in the floodplains, would kill the other half—that is, kill the Black River.

When not flooding, the Black River was a quiet, shallow waterway, shaded by dense vegetation, and rich in sediment washed out of the Cedar's old river terraces, hence its name, as opposed to the clearer White River, which it joined to become the Duwamish.[9] What the White lacked in color it made up for with spring floods that pushed up the Black and reversed the darker river's flow back into Lake Washington. This is the origin of the name for the Black in Chinook jargon: Mox La Push, or "Two Mouths." Modern roads and buildings mask how flat the White River valley is and how it easy it would have been for a large flooding river to rage across the level land and up its smaller tributary.[10]

Despite its modest length, the Black had an oversized role as the only outlet of Lake Washington, notes historian David Buerge.[11] He describes the land surrounding the Black as a "storied landscape," where the hills, the river, and a quaking bog were associated with supernatural beings and events. For the Native people, living along the Black meant controlling access to the salmon that used the river to reach the lake and the Cedar River. This location made them "well-to-do and well connected," with three major settlements along the Black and an extensive trade network down the river to Elliott Bay and Puget Sound.

With the building of the ship canal, no fish would ever again travel the Black River, though traces of the channel persisted for decades. A photograph from 1969 shows what was apparently the river's last remnant in town, a narrow swath of shrubs, weeds, and cottonwoods in downtown Renton. When I go to find that historic waterway, I become excited when I discover a row of trees at roughly the right location. But when I examine the photograph with my reading glasses on, I realize I am too far east, and that the last vestige of the Black in Renton now lies

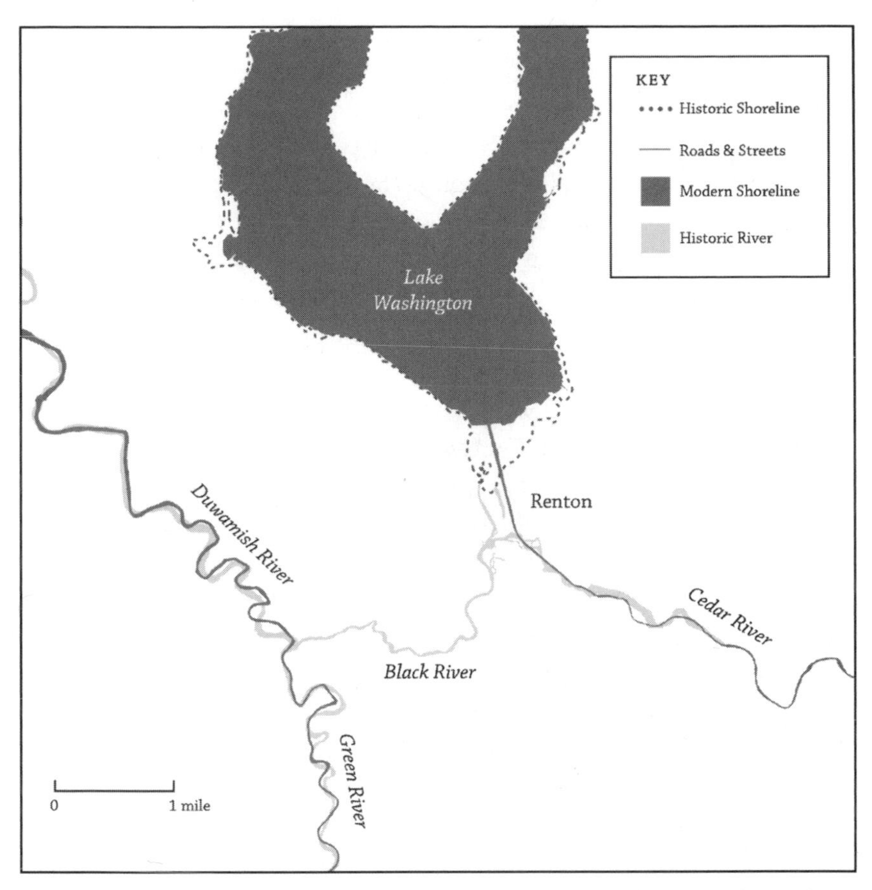

MAP 4.3. DETAIL OF BLACK AND CEDAR RIVERS. Historical mapping by UW's Puget Sound River History Project.

under the parking lot of a huge Safeway. But one clue still remains, two blocks west of the Safeway, behind a Fred Meyer store.

The gentle arc of Hardie Avenue Southwest follows what was formerly the route of the Black River as it meandered south. Farther down Hardie, the road continues under a set of railroad tracks, as the Black once did. The river then turned west and flowed south of the tracks to its confluence with the Duwamish River. As far as I can determine, Hardie is the only infrastructure left that provides a trace of the Black.[12]

But like Westley in *The Princess Bride*, the Black is not all dead. Over the past decade or so, a dedicated group of volunteers has worked to restore habitat along a short expanse of water, part of an area now called the Black River Riparian Forest and Wetland. It is about a mile west of Hardie and less than a mile north of Southcenter Mall. Crews have removed exotics, mulched the soil, and planted native plants, and this has resulted in visitors reporting over fifty species, including salmon, coyotes, salamanders, and bald eagles. The area also supports the largest heron rookery in the region.

When I walk down to the water through the lush plants, I find it hard to imagine salmon surviving, much less swimming, in it. The water looks fetid, brown, and skanky, with a shopping cart plopped down in the center of the narrow channel. Perhaps the rainy season will lead to clearer water and someone will remove the metal cart.

Returning to downtown Renton and the land of concrete and industry, I am reminded of two observations about the Black and its fate. In referring to the soon-to-be-dried-out Black River, Hiram Chittenden wrote that the best use of its channel would be "industrial purposes."[13] To people such as Chittenden and Renton's civic leaders, reengineering nature was simply what needed to be done to improve the local community. They were eliminating flooding, creating space for a new port, and opening up access via the ship canal to a much wider world of trade. Fixing the Black and Cedar was about the future; those rivers and their ways were the past.

Joseph Moses, who was a member of the Duwamish tribe and one of the last to live on the tribe's historic land in downtown Renton, offered a different take. "That was quite a day, for the white people, at least. The waters just went down, down, until our landing and canoes stood dry, and there was no Black River at all. There were pools, of course, and the struggling fish trapped in them. People came from miles around, laughing and hollering and stuffing the fish in gunny sacks." In the words of Warren King George, the Black River is like "a bad scar that never heals up."[14]

◆ ◆ ◆

Redirecting the Cedar and killing the Black did more than simply eliminate flooding. The changed hydrology altered the lifestyle of the fish in the lake. For example, chinook and coho salmon, and probably steelhead as well, had to ferret out the new exit from the lake via the locks.[15] In contrast, longfin smelt did not or could not make this change. Biologists suspect that these four- to five-inch fish became isolated in the lake and adapted to spend their entire lives in freshwater. All other populations of longfin smelt found along the West Coast, except two isolated communities in British Columbia's Fraser Valley, migrate between saltwater estuaries and freshwater streams. At first glance this might not appear to be environmentally significant, but biologists have proposed that the year-round presence of longfin smelt, as opposed to those that historically would have been transient, played a key role in helping to improve the lake's clarity in the 1970s.

In the late 1950s, Lake Washington achieved unwanted notoriety because of the abundant sewage dumped directly into the water. This effluent contained a high concentration of nutrients that plankton fed on, which led to an explosion of cyanobacteria known as *Oscillatoria rubescens*, making the lake unpleasant, unhealthy, and murky. In response, voters approved the establishment of the Municipality of Metropolitan Seattle, or Metro, to build and operate regional wastewater treatment facilities that diverted sewage away from the lake and into Puget Sound. By 1968, 99 percent of the diversions had been completed and the subsequent crash in the *Oscillatoria* population was hailed as a successful recovery of the lake. Then in 1976 Lake Washington surprised everyone and became more transparent than ever before.

Biologists zeroed in on a one-eighth-inch-long crustacean called *Daphnia*, an animal known for its ability to keep lakes clean via a diet of smaller critters and particles. In 1976, the lake's *Daphnia* population exploded. Researchers initially proposed two reasons. The decrease in *Oscillatoria* was key because the cyanobacteria inhibited the *Daphnia*'s ability to eat. They also noted a decrease in another crustacean, *Neomysis mercedis*, a major predator on *Daphnia*. But why in 1976? The answer was longfin smelt.

Although the Cedar River no longer flooded through downtown Renton, it still flooded upstream and carried huge amounts of sediment downriver, which required yearly dredging. After an exceptionally large flood in 1959, King County changed tactics and decided on a multiyear project to build revetments, or piles of large boulders, along the shoreline to reduce bank erosion during high flows. Doing so meant that fish-spawning beds in the river were no longer destroyed by dredging. One fish that took advantage was longfin smelt, whose preferred food choice was *Neomysis*. The smelt appear to have been the tipping point in the community, reducing *Neomysis* populations to low enough levels to allow *Daphnia* to thrive and cause the lake to clear.

But a change in biology does not tell the entire story. Because the Cedar River drains an upland forest, it contains relatively low amounts of nutrients, particularly compared with the lake's previous main source, the Sammamish River. When the sewage discharge was stopped, water from the Cedar helped to quickly dilute the nutrients in the lake water. In addition, the Cedar carries significantly more water than the Sammamish, which has the effect of increasing how quickly the lake water replaces itself. Before 1916, the residence time of water in the lake was five years, about twice the present rate. The twofold effect of dilution and fast flushing helped clear the lake more quickly than anyone had predicted.[16]

Not only did the animal community change but so did the plants. Historically, hardstem bulrush, or tule, another key plant collected by Native people, thrived along the shoreline. At present, it is hard to find any tule stands. Several reasons account for their loss and that of other emergent species, or plants that grow in water. One is habitat loss. A 2001 study reported that 2,737 docks bordered Lake Washington, an average of 36 per mile. In addition, nearly three-quarters of the shoreline is either riprap or bulkhead. Before 1916, trees overhung the water, slopes slid into the lake, and emergents flourished along the shoreline. Visit the lake now and mostly what you'll find are cement walls, big boulders, extensive docks, and grassy slopes.

The second challenge for shoreline plants is what University of Washington ecologist Si Simenstad calls the "upside-down seasonal hydrology."[17] As one Harold Smith told Lucile McDonald, the lake was highest in winter. For Smith, who grew up in the 1880s around what would become Seward Park, this meant that the peninsula "often was an island and the isthmus could be crossed in canoe or rowboat," with the lake regularly rising six or seven feet higher than during the summer low period.[18] With the lowering of the lake, the land reverted to permanent peninsular status. (Dropping the water level also changed the status of the formerly isolated Pritchard Island and Foster Island; both could now be reached by a land connection.)

Since 1916, the lake level peaks in the summer, no more than two feet above the winter low. In contrast to the natural rhythm, when snow and rainfall controlled the level, the Army Corps of Engineers now regulates the lake level. They do so because too much fluctuation would not be good for the houseboats and moorings on the lake or for the floating bridges. If the lake dropped too low, the cables holding the bridges in place would become slack and lose their integrity.

For plants adapted to a winter high, a summer high can be catastrophic. Emergent plants typically set seed in the autumn and germinate that year or the following spring. During the modern hydrologic regime, the plants don't have enough time out of the water in the summer, when they need to get established and grow. Add to this ecological challenge the secondary effect that these plants, adapted to relatively large fluctuations, must now try to grow in an environment that is more or less static, and you have a recipe for a less healthy ecosystem. "Lake Washington is essentially a poster child for how we can disrupt a system. It's fairly remarkable that it's not sterile with all that we have done to it," says Kurt Fresh, a biologist with the National Oceanic and Atmospheric Administration.[19]

The biggest habitat loss around the lake has been the wetlands and shallow-water areas, such as at Renton. They were critical ecosystem losses. Wetlands helped filter sediments out of the streams that washed into the lake, acted as a sponge absorbing excess water, and protected the shore ecosystem during severe weather. They also provided good habitat for birds and fish.

Along with the marsh at Renton, the largest wetland to disappear was at the north end of Union Bay. My map shows that, in 1905, Ravenna Creek drained out of Ravenna Park and into a wetland, or bog, that spread across what is now University Village. Also shown is the old Seattle Lake Shore and Eastern Railway, owned in 1904 by Northern Pacific and now known as the Burke-Gilman Trail. The tracks follow the land contours and wrap around the bog instead of taking the direct route across. McDonald adds that the train passed by the town of Yesler, with its one-room schoolhouse, two churches, and a post office.[20] Platted in 1888, Yesler was home for the families who worked in the mill Yesler had established on the shore of Union Bay. In 1892, he employed thirty-six men, who cut seventy-five thousand board feet of lumber every twelve hours, which was then put on railcars and taken back to Seattle via a spur from the main SLS&E tracks. North of the town, the train headed through what McDonald called "wild country."[21]

Like most local bogs, the Ravenna bog had pockets of open water interwoven with a quagmire of floating mats of decomposing plants,

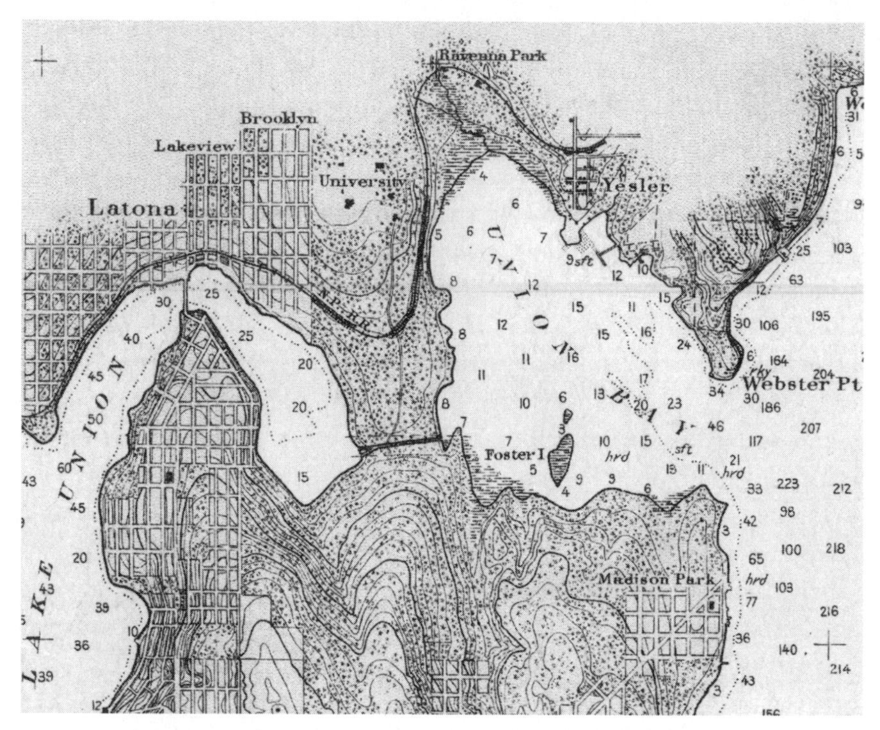

FIG. 4.2. LAKE WASHINGTON TOPOGRAPHICAL MAP, DETAIL OF UNION BAY, 1904. Courtesy University of Washington Libraries, Map Collection.

topped by ground covers such as cranberry and sundew—one of the state's few carnivorous plants—and an overstory of bog laurel and Labrador tea. But with the lowering of the lake and a consequent altering of the Ravenna Creek drainage, the bog began to dry out.

Truck farmers—mostly Japanese and Italian immigrants—were the first to exploit the bog's thick soils of nutrient-rich peat. They began by scalping off the top foot or so, hoeing the upturned soil into wheelbarrows, and carting it away over a series of planks. They would then loosen the new surface with spades and let it aerate for a year or two before planting it with vegetables.[22] To sell their produce, the farmers ran what could be called a child's nightmare: an ice-cream-truck-like operation, but one that sold spinach, cabbage, and beets instead of frozen treats.[23] The lowland remained a favored spot to grow plants until 1954, when the Malmo Nursery sold the land to developers, and Seattle acquired its second mall, University Village. Coincidentally, Seattle's

Fig. 4.3. UNION BAY, CIRCA 1916. Before the lowering of Lake Washington, Union Bay was actually a bay and extended north to a point near present-day Northeast Forty-Fifth Street. In 1914, the city built a wood trestle on Forty-Fifth Street across the upper end of the bay, to Laurelhurst. Unfortunately, several vehicles plunged off the narrow roadway and into the swamp below. Courtesy University of Washington Special Collections, UW 36353.

first mall, Northgate, about three miles northwest of the Village, was also built atop an old bog.

To the south of the bog lay Union Bay. A photograph taken in 1916 gives a feel for the calm of the romantic inlet. The photographer looked east toward the recently logged Laurelhurst peninsula. On the far side of the open water are Yesler's mill, several floating buildings, a variety of pilings, and a boat dock on the peninsula. On the near side is a mix of small boats and the University of Washington boathouse and dance hall. A line drawn across the bay from the white-roofed boathouse to the cluster of floating structures on the east side would mark the modern boundary of land and water. This north extension of Union Bay had been a very popular spot for University of Washington students, for what was known as "canoeing wooing." On the west side of the bay, where the modern IMA building now stands, the students had built a boathouse with a dance hall and racks for rental canoes to further their amorous pursuits.

When the water drained out of the bay in 1916, a cattail marsh emerged south of Northeast Forty-Fifth Street and the former bog. Little appears to have happened in the marsh during the early years after the opening of the canal, but in 1926 the city decided that the marsh would serve better as a dump. Waste disposal began at the northeast corner, adjacent to the five-corners intersection. Until the dump closed in 1966, a cover of industrial and household materials slowly devoured the marsh. Initially garbage had been burned in open fires, which resulted in "many justifiable complaints from nearby homes." Beginning in the 1950s, new garbage was covered daily with soil.[24]

By the late 1950s, trucks were bringing an estimated 110 loads every day, belching up between 40 and 66 percent of Seattle's refuse. The city's trash, in places forty feet deep, weighed so much that the peat underlying the garbage began to creep into Union Bay, which led to the construction of containment dikes made from used telephone poles and rubble. The trash also attracted thousands upon thousands of gulls, making it an astoundingly noisy spot. When the dump closed, the entire two hundred acres of the former marsh had been filled with trash and dirt; it is now covered by parking lots, a driving range, storage yards, playfields, and the Center for Urban Horticulture.

In the more than forty years since the first attempts at restoration, the area around the Center for Urban Horticulture, known officially as the Union Bay Natural Area and unofficially as the Montlake Fill, has become a true treasure for visitors, who find solace, beauty, and a rich diversity of wildlife. They have sighted more than two hundred bird species, as well as beavers and salamanders, cottonwoods and willows, and dragonflies and butterflies. The area is not only a refuge but also an inspiration for how to restore damaged land.

Sanitary landfills, as these garbage dumps were euphemistically labeled, dotted the Seattle landscape up through the 1960s. (The other one on a former Lake Washington wetlands was located at Wetmore Slough, or modern-day Genesee Park.) They were seen as a way not only to aggregate trash but also to reclaim and improve land.[25] At least a dozen were placed in low-lying areas or ravines, usually owned by the city but sometimes by private citizens who wanted to raise their property to street level. The dumps took both putrescibles and nonputrescibles, though few records exist that document which potentially toxic industrial materials also went into the landfills. Burning was common, rats were abundant, and the odors were abhorrent. And like the Union Bay dump, these dumps were eventually covered over—and often forgotten, or at least concealed so they would not remind us of how we treated the land—and converted to more practical uses such as parks and building sites. A report from 1984 estimates that up to two hundred buildable acres had been added to Seattle by converting wetlands first to rubbish and then to filled land.[26]

◆ ◆ ◆

For promoters of the ship canal, the loss of the lake's wetlands was not problematic. In fact, it was seen as a benefit because the "drainage of swamps will improve rather than detract from the natural beauty of the shore line," wrote one engineer working on the project.[27] He added that settlement of the newly exposed land would also lead to planting of turf and shrubbery; or better yet, the land would be an ideal location on which to build factories, which many Seattleites had long wanted around the lakes.

One such factory was the *P-I*'s impetus for predicting that the Seattle area would be the "Pittsburgh of the Pacific Coast." In 1886, Peter Kirk, owner of the Moss Bay Hematite Iron and Steel Works in England, had been attracted to the area because of Thomas Burke and Arthur Denny's reportedly rich deposits of iron ore near Snoqualmie Pass, nearby beds of limestone (needed as a flux in steel production) and coal at Newcastle, and the growing rail system. Kirk returned two years later and teamed up with several influential Seattleites, including Denny and Leigh S. J. Hunt, publisher of the *P-I*, to form the American branch of Moss Bay. Soon, announced Hunt's paper on May 31, 1888, a town of ten thousand people would arise on the shores of Lake Washington. In preparation for the grand scheme, the company built a brickworks and sawmill, began construction on the iron mill, and developed a model town for its new employees. But then the nationwide Panic of 1893 hit, outside sources of money disappeared, and Kirk's steel-making future—and that of his eponymous town—died like so many other grand plans in Seattle. McDonald wrote that investors lost more than a million dollars.[28]

So when the lake dropped, entrepreneurs were ready; but instead of industry, sunbathers and swimmers exploited the land newly exposed by lowering the lake. Their favorite place to go was Juanita Beach Park, north of Kirkland. Before 1916, Lake Washington came all the way up to Juanita Drive, just below where an early settler, Dorr Forbes, had dammed a small stream to create a pond for a shingle mill. Forbes's son Leslie told McDonald that he remembered seeing Indians canoe into the bay "right up until the time the lake was lowered."[29] They had come to gather wapato, a potato-like tuber that grew in the shallow water. The bulbs were a key element of their diet and were harvested in the fall. Across the lake was a spot known as Digging in the Water, a reference to the importance of wapato to the Native people.

Forbes's mill was long gone by 1916, but the family still owned the land. When the lake dropped, their property grew out into the lake, where a beautiful sand beach had appeared and the wapato had disappeared. Word got out about the beach, and people began to arrive, which prompted Leslie to clear the land of old logs and cattails. To make the area more attractive, he planted cottonwoods, although roaming cows regularly destroyed them. In 1921, Forbes opened Sandy Beach to the public and charged a small fee. The only amenities were two outhouses—or Chic Sales, as Forbes's daughter Dorris Beecher called them—followed in later years by a bathhouse with dressing rooms and a dance floor, a two-plank walk to the beach (because people complained about sand clinging to their feet), an open-air kitchen, and picnic tables.

The Forbes family didn't make a lot of money, but the beach did sustain them, despite the best efforts of some patrons. When the family charged twenty-five cents per picnic table, people began to sublet them, making more money than the owners. When the Forbeses charged twenty-five cents per car, people parked outside the park or loaded everyone into one car. Finally they charged one dime per person. Visitors could also rent swimming suits, black with red stripes. The thickness of the stripe indicated the size. No one came up with a way to sneak around this charge.[30]

Business peaked in the 1930s, when several thousand people would convene on the beach on weekends. Sandy Beach lost its prestige after World War II, in part because the easy access to the east side of the lake provided by the recently opened Lake Washington floating bridge made the area less exotic. The original owners sold the property to King County in 1956. A year later the county purchased the adjacent beaches and combined them into Juanita Beach Park, where you can still find a sandy beach, a place to swim, and picnic tables, all for no cost.

Not every pleasure pursuit benefited from the lowered lake, wrote McDonald. In 1897, Captain Ferry Burrows had opened his Summer and Winter Pleasure Resort at the mouth of the Black River. The resort attracted sportsmen from the north—that is, Seattle—who would ride the street trolley to Rainier Beach, where Burrows met them and rowed them back to the boathouse for a weekend of fishing, hunting ducks on and around the Black River, and eating Mrs. Burrows's chicken and biscuits. Unfortunately for the Burrowses and their resort, the disappearance of the Black put them out of business and the Burrowses had to move into town.

◆ ◆ ◆

But industries did develop on the lake. Most were drawn there not by the newly exposed land but by the new connection out to Puget Sound and the Pacific Ocean. Perhaps the oddest to appear was one established on Meydenbauer Bay, now one of the eastside's more exclusive neighborhoods but formerly the winter base for what would become the last substantial whaling company based in the United States. Owned by William Schupp, who had made his money in insurance, the American Pacific Whaling Company had long been based in Westport, Washington, then known as Bay City, when Schupp acquired it in 1914.

In 1918, he moved his fleet to Meydenbauer Bay to take advantage of the less damaging freshwater. The seven killer boats, as they were known, would overwinter from October to April for maintenance and repairs, most of which took place just before departure north. Although no processing of whale products took place, Schupp's decision to move to Bellevue added significantly to the economy. To give an idea how small the town was even in 1941, the famous Works Progress Administration guide to the Evergreen State noted only that Bellevue was "a junction with a concrete-paved crossroad," where "hundreds of pounds of shortcake are served to visitors" at the annual strawberry festival.[31] In contrast, Schupp's whaling fleet merited two paragraphs and song lyrics.

FIG. 4.4. KILLER BOATS, MEYDENBAUER BAY, 1937. The largest of the seven killer boats of William Schupp's American Pacific Whaling Company were the 105-foot *Kodiak* and *Unimak*. Eleven men typically operated each ship, with about a hundred people working at the Meydenbauer Bay facility in winter. Courtesy Eastside Heritage Center.

Along with the boats came the whaling crews, mostly Norwegian men, some of whom spent the winters around Bellevue and some of whom overwintered in the less respectable parts of Seattle. Schupp's grandson Bill Lagen recalled in a letter to Lucile McDonald that, in the spring, his father would ask the police to start rounding up the men about two weeks before it was time to head to Alaska and to "keep them in the tank at the old city jail on Yesler." Bill and his dad would then drive down in a big Packard and collect them. To find those not incarcerated, Bill, only twelve years old or so, would search the bars. "There would be a whoo, and a 'Hello, Bill,' and we would go out to the car and collect another load in various stages of sobriety, which we would deliver to the ship, and then away we would go for Alaska."[32]

With concerns about Japanese attacks in Alaska during World War II, the navy leased the whaling boats for patrol use. After the war, the family attempted to start whaling again; but when Schupp and his son died, the company did, too. In 1949 the Sealand Construction Company bought the killer boats, leading to a fitting end to Bellevue's whaling era. Sealand took the boats to the Duwamish Shipyard, where each was stripped, scrapped, and reduced down to its constituent elements, as if it were a whale and the sum of its parts was viewed as more valuable than the whole. After the war, the Lagen family helped develop the marinas now found in Meydenbauer Bay.

◆ ◆ ◆

The largest industry to develop, or at least expand significantly, after the opening of the canal was boatbuilding, one of the oldest commercial industries on Lake Washington. In 1886, Edward Lee opened a shipyard on Pontiac Bay at the northwest corner of present-day Magnuson Park, or Sand Point as it was also known. Lee's shipyard produced boats only for local passenger service.

For much of the history of Seattle and the communities on the east side of Lake Washington, the best, and often only, means of travel was by boat, either self-propelled in rowboats, or via steamboats and paddle-wheel ferries. Most of these boats were locally made or retrofitted. At least eighty-five steamers crisscrossed the lake over the years, functioning as the equivalent of our modern bus system and running regular service to designated spots, usually termed *landings*, either officially named ones or simply docks where people lived.[33] At smaller landings, passengers waved a white flag to get the boat to pull in. In fog, they could ring a bell or old saw blade, though McDonald talked to one old-timer who told of a dark and stormy night when the boat missed the dock and ran aground in their yard. Another historian told me that the boats would stop for women, but men would have to leap onto the boat as it passed the dock.[34] Tickets ranged from a dime to a quarter.

Those taking the regular ferry service did not need flags or dexterity. For example, trips to the east side of Lake Washington had scheduled service from a dock at the east end of Madison Street. During peak times, Seattle residents could catch trolleys that left downtown for Madison every two and a half minutes; as in modern times, people were careful not to miss the last boat or trolley, which departed at 1 A.M. One correspondent of McDonald's recounted a February 1887 trip to attend a wedding at Lake Sammamish. The young woman caught a boat at Madison, which took fourteen hours to reach the upper lake. After the wedding, the bride's father took his wife and the young guest by canoe to Redmond (near the present-day town center), where the women would catch a stage to Houghton, on Lake Washington's east side. Unfortunately it never came, so they walked four miles in the snow back to Houghton, where the next day they caught a steamer back to Madison.

One slow-to-appear consequence of lowering Lake Washington by nine feet was how the newly exposed land altered this historic means of transportation. For example, before 1916 there was just enough room along the bluffs north of Sand Point to accommodate one set of railroad tracks, which is one reason why ferries were necessary. Elizabeth Stewart, director of the Renton History Museum, offers several other reasons for the lack of roads: low population, the lack of interest in spending money on paved roads to move people (because early property and business owners were taxed to pay for roads that cut across their land), and the efficiency with which railroads and electric trains moved goods to market (moving goods was, until the 1920s or so, more important than moving people to population centers).

After 1916, though, enough newly exposed land appeared that roads could be added; there was also room to build the houses that now dot the lakefront, which accounts for the houses below the Burke-Gilman Trail along the northwest edge of the lake. But those homeowners have probably learned that the steep bluffs that rise above the trail are prone to landslides, a problem exacerbated by the fact that the SLS&E cut the base of the slopes to create space for tracks. By 1925, enough new roads had been built to severely cut into the ferry traffic, which limped along to its final service in 1950.

With the opening of the canal, the maritime industry turned to building boats for distant clients. Anderson Shipyard, the best-known boatbuilder, was at what is now a development of shops and offices known as Carillon Point, where boatbuilding had begun in 1901. Unfortunately, the first boat built at the point got stuck for months at the mouth of the Black River, so for the next seventeen years all boats built at the point, primarily by Anderson, stayed in the lake.

Business exploded when Anderson was able to send boats out into Puget Sound, just ten miles away via the ship canal route. Their first customer, the New York–based Oriental Navigation Company, commissioned two wooden 270-foot ships. When the first one, the *Osprey*, launched in 1918, more than five thousand people watched the festivities, which included speeches by Thomas Burke and former territorial governor Watson Squire, who remarked, "Today we behold the first fruits of the canal."[35]

Those fruits continued to flourish up through World War II, when six thousand men and women worked at the shipyard. There was so little housing around the point that many employees lived in Seattle, which led to a dedicated ferry service, direct from Madison Park to the plant. The shipyard had played a critical role in the development of Kirkland; but with the end of the war, the era of the eastside shipyards began to die. They were replaced over the decades by marinas that catered to a different boating lifestyle. Now, the main evidence of the extensive Anderson Shipyard rests underwater, where divers have identified at least thirty wrecks, old pilings, the launch ramp for the ships, and a scattered bounty of lines, cables, and blocks.

Wandering through the modern spas, fitness center, cafes, and artist's cooperative at Carillon Point, I find it hard to imagine the humble development drawn on my map. It shows that between Juanita Bay and Bellevue, there were no more than seventy-five houses and one main road with a few spurs. With the opening of the ship canal, though, the eastside would no longer be a quiet, way-off-the-beaten-path place. The new connection to Puget Sound and the Pacific Ocean was the first step in making what had been Lake Washington's quaint eastern shoreline communities into the modern cities of today.

◆ ◆ ◆

But the lowering of Lake Washington did not aid all industries equally. Long before the canals, timber had been the primary source of jobs and income for those who lived around the lake. McDonald describes mills at Yesler, Kenmore, Juanita Bay, Renton, and Leschi and refers repeatedly to logging along the shoreline and inland. By the early 1900s, many mills were on the decline, but one of the largest persisted during canal construction at an unlikely location: more than two miles north of the modern shoreline of Lake Washington. The spot is now roughly where Interstate 405 passes under Main Street, about a mile southeast of downtown Belle-

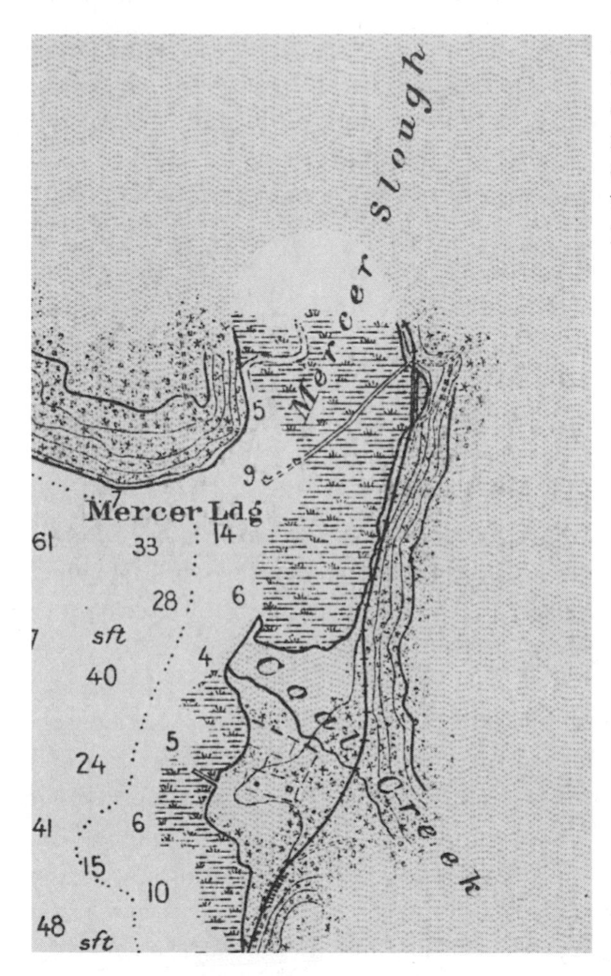

FIG. 4.5. LAKE WASHINGTON TOPOGRAPHICAL MAP, DETAIL OF MERCER SLOUGH, 1904. Courtesy University of Washington Libraries, Map Collection.

vue. From the 1890s to 1916, boats up to fifty feet long regularly traveled up a former arm of the lake—the Mercer Slough—ferrying freight, hunting and fishing parties, and passengers, many of whom lived in the town of Wilburton, the end point of the waterway. On the hill next to Interstate 405 is the town's defunct train stop. All that is left of the former arm of the lake is a waterway that has been reengineered and narrowed to the point that the only way to explore it is via canoe and kayak.

The timber industry started in 1894 with a small dam at the upper end of the slough. A pond provided storage for logs, which were floated through the slough to the lake. Ten years later, the Hewitt-Lea Logging Company acquired the property and platted Wilburton for its workers. They also built a mill and a formal wharf for loading wood onto boats.

During the dozen years they logged the eastside forests, Hewitt-Lea took out 100 million board feet of lumber, 1 million feet of small cedar logs, and twenty-four thousand to thirty-six thousand pilings.[36]

Access to the wharf and mill ended, though, when the ship canal opened and the water level dropped too low for boats to travel the two miles from lake to wharf. In response, Hewitt-Lea sued King County, which controlled the right-of-way in the ship canal, for $125,000 in damages. King County Superior Court rejected the case on the grounds that building the locks improved the navigability of the lake, and that whatever losses Hewitt-Lea sustained were incidental to the new and improved lake navigation. The company appealed to the state supreme court, where they won a retrial. In 1924, the superior court ruled against Hewitt-Lea again, in part because the county made the case that the value of the mill site had already declined because loggers had stripped the forest before Mercer Slough dried out.

After the loggers moved out, flower farmers moved in. The modern landscape reflects the later development of the area, particularly the upper end surrounding an office park, where developers dredged a looped waterway around part of the property. The owners also donated land to the city for the present-day park. Though still a wild spot in the heart of suburbia, the present Mercer Slough is more a product of human manipulation than of nature.

◆ ◆ ◆

Not only was logging on the way out but so was coal. By the time McDonald was writing her articles, she could note that names such as Factoria and Coal Creek "must puzzle newcomers," since few people realized the importance of coal to the eastside.[37] Those names seem even more bizarre today, with nearly every vestige of the coal industry erased. The few reminders are features such as the coal adits, concrete foundations, and old railroad grades that dot Cougar Mountain Regional Wildland Park, but those fail to truly convey the importance of coal to the Seattle region. The same situation is true with logging: one of the few acknowledgments is Log Boom Park at Kenmore, which honors the great log rafts, each of which was secured with a boom held together by cables.

What these changes reflect is a wholesale movement away from the dependency—of Seattle and communities around Lake Washington—

upon the natural-resource extraction business. The towns had grown on the raw goods they could supply to the outside world, but much of the lumber was gone and coal was no longer in demand. Industry had to, and did, change in response to the new reality; this is evident in the development that occurred around the lake after the ship canal was built and has continued up to the present.

This stands out at the end of a six-hundred-foot dock that reaches into the middle of Salmon Bay. One of the main centers of Seattle's modern maritime industry, with several shipyards and the Port of Seattle's Maritime Industrial Center, the bay is also the home base for the North Pacific fishing fleet, which annually brings about $5 billion to the Seattle economy. The fleet is the modern equivalent of the old whaling boats: it goes north to harvest the abundant marine resources and returns here to overwinter in the freshwater and benign weather.

At this industrial hub, I am struck by the differences between the three bodies of water that the locks and ship canal connected. In contrast to Salmon Bay, Lake Union, which initially developed as a boatbuilding center, is now ringed by houseboats, marinas with private boats, and a scattering of industrial structures. Ironically, its other big industry (the production of synthetic natural gas) and the one military facility (Naval Reserve Center) that formerly thrived on Lake Union have been converted to two green spaces, Gas Works Park and Lake Union Park, respectively. Residences are the dominant feature on Lake Washington's shoreline; a small percentage of industrial or commercial concerns are located on the water. On Lake Washington, in contrast to Lake Union and Salmon Bay, green spaces and beach constitute roughly 30 percent of the shoreline. The lake also has many more areas of public access.

Of the three bodies of water, Salmon Bay is the only one that fits the industrial vision of those nineteenth-century Seattleites who fought for a ship canal. Industries did develop on the shorelines of both lakes, but not really to the extent hoped, and no navy port ever materialized. This raises the question of whether we could have simply left the situation the way it was, with three separated bodies of water. After all, Ballard had a good industrial base in Salmon Bay before the locks were built. Or could we have connected only Salmon Bay and Lake Union, without extending the ship canal to the bigger lake? Hydrographically this could not have happened. In order for the locks to operate, they need the volume of water supplied by the Cedar River and Lake Washington. Not enough water was in the historic watershed of Lake Union for locks to have functioned, but that did not preclude the possibility of dredging a wider canal or building some other connection between Lake Union and Salmon Bay that did not involve locks and thus did not need an additional source of water.

But did the locks and ship canal have to be built? They should not have been built, from an ecological perspective. An optimist might cite the role of the smelt in helping to clear up Lake Washington as a benefit, but that was purely unintentional. Others might point to the reduction in floods on the Duwamish—a result of killing the Black River—as a positive outcome, but the historic flooding aided the native fish, who could exploit the habitat of the flooded wetlands on the Duwamish. One reason flooding on the Duwamish was problematic is that people eliminated the wetlands, which were able to absorb the excess water and lower the impact of floods. There can be no doubt, though, that destroying the Black River and its role as a migratory corridor outweighs any other perceived ecological benefit.

The economic question is a bit muddier. Boatbuilding was the key industry that developed because of the locks, not only in Lake Washington but also in Lake Union and Salmon Bay. Could this have occurred without the locks? Certainly not in the lakes, but possibly in Salmon Bay. Landowners around Salmon Bay could have adapted to make room for the industries that developed on the lakes, though it would have been much more challenging with the bay's tidal fluctuation and the salt water. In addition, without the bay's conversion to freshwater, the North Pacific fishing fleet probably would not have the presence in Seattle that it does today.

Perhaps the best reason for building the locks and canal was the one stated by Thomas Mercer, who said their purpose was to "provid[e] a connecting link uniting the larger lake and Puget Sound."[38] At its most basic level, his comment is about transportation in our glacial landscape. Mercer recognized that the unlinked bodies of water were comparable to the hills and bluffs that made travel around the downtown corridor a challenge. If you could build an easy-to-use transportation network between salt water and freshwater, you would eventually open up new areas for people to use.

If you have any doubt about how Mercer's prophetic statement has played out, go to the modern locks. The most common sight is pleasure craft used by people who see the canal and locks as a travel corridor. In this regard, the locks are one of the best lines of evidence of Seattleites' will to shape the city's topography to their needs. In watching boats move between fresh- and salt water, we can sense that the locks represent change. By definition, that is what a set of locks indicates. Something was not right, not usable, or not practical and had to be fixed. In Seattle's case, Lake Washington and Lake Union could not be fully utilized for transportation and industry, because the lakes didn't have good connections to salt water. By building the locks, Seattleites aimed not only to integrate the lakes more fully into the city's economy but also to make it easier for the rest of the world to connect with Seattle.

Improved transportation corridors ultimately became a fundamental way of envisioning a better Seattle and an idea that acted as a central driver for reshaping our topography. Improved transportation was the reason for the decades of railroad and tideflat-filling battles along the waterfront and is still a primary reason for how we view our topography, as illustrated by our constant struggles over transportation funding and projects such as the State Route 99 tunnel. Enlarging and improving what he called the city's arteries was also a main justification cited by the man responsible for Seattle's most dramatic topographic change. To Reginald Heber Thomson, a functioning city was like a human body; and according to him, "just as a free circulation of blood is necessary to secure the development of each part of the body, so a free movement of traffic is necessary to develop each portion of the city." Continuing the metaphor, he emphasized that regrading projects tended to "enlarge insufficient arteries and to remove stoppage or blockades which heretofore existed in those arteries tending to prevent a free flow of the blood."[39] And no blockage in Seattle was worse, in Thomson's eyes, than Denny Hill, or what he called "an offense to the public."[40]

H.W. HAWLEY.

5

Regrading Denny Hill

LIKE MANY OTHERS IN THE EARLY 1880S, R. H. THOMSON CAME TO Seattle because of its great potential. With abundant coal and timber, and the impending expansion of train service into the city, Seattle was a perfect opportunity for a young man—or so a family friend, an engineer for the Northern Pacific Railway, told Thomson. "This constant strain of kindly advice wore on me," wrote Thomson in his autobiography, so on September 20, 1881, he got on board the side-wheel steamer S.S. *Dakota* for the five-day trip to Seattle from San Francisco.[1] At the time, he had had a short teaching career in California and had made several unsuccessful stabs at mining engineering.

He hoped to change his luck in Seattle. Born and raised in Hanover, Indiana, Thomson had taken surveying classes in college but otherwise had little professional experience when he landed in the city. He did have a cousin, Frederick "Harry" Whitworth, who was the city surveyor for Seattle. Whitworth also had a private practice and hired Thomson to join him. Working for his cousin, Thomson got to know the land in and around Seattle while surveying terrain for the railroads, property owners, and the local coal mines. In August 1884, one year after his cousin Harry quit the job, Thomson became the city surveyor, a position equivalent to that of city engineer.

Thomson's "greatest triumph" during his early years, writes his biographer, William Wilson, was building the Grant Street Bridge.[2] This wooden trestle ran out over the tideflats along the base of Beacon Hill and would become part of a major transportation route south of the city. The

Watching the regrade, 1907 (detail)

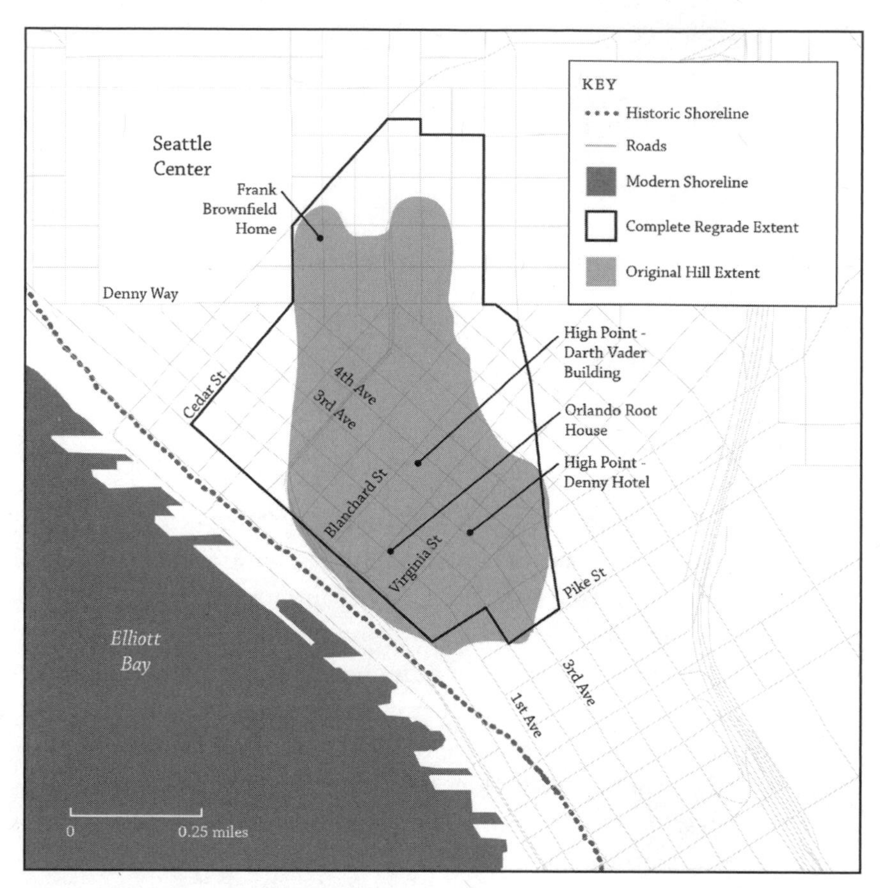

MAP 5.1. DENNY REGRADE SHOWING BOUNDARY OF HILL AND BOUNDARY OF REGRADE WORK. The outlines of Denny Hill are based on areas defined by city ordinances and topography. Topographic data courtesy Aaron Raymond.

structure exemplifies a central theme of Thomson's: connecting disparate parts of the city. As he later wrote in his autobiography, the extremely hilly landscape made him wonder, "How will the people in one end of the city be able to do business with those in the other end?"[3] The answer was to regrade what was in the way and open up corridors that would allow people to get where they needed to easily and quickly.

Seattle, noted Thomson, was particularly in need of this kind of fix. The problem was that the city's founders had platted the land "with but little regard as to whether the streets could ever be used or not, the main idea being, apparently, to sell the lots."[4] What they should have done, instead, was to lay out the trails and roads first. Not that Seattle was alone in making this error. "In the ruins of very ancient cities, we find evidence that the same character of work—that is to say the work of regrading—was carried on before the beginning of history," wrote Thomson.[5] Now it was Seattle's time to carry out this ancient process, and he knew the perfect place to put his ideas to work—the offensive Denny Hill.

◆ ◆ ◆

When the Puget lobe glacier retreated to the north around 16,400 years ago, it left behind a bald rise perhaps better described as a narrow wedge than a hill. The low mound trended north-south, rising gently to two summits, both of which topped out at about 240 feet above sea level. At the south end was the hill's steepest face, which gazed out over what would become Seattle's central business district. On the east the hill sloped to a valley and a small stream that carried water north to Lake Union. The west slope was almost as steep as the south face and dropped to a small flat before dipping steeply into Elliott Bay. Of the seven hills that Seattleites claimed as their home territory, this was the shortest, dwarfed by Capitol Hill to the east and Queen Anne to the northwest, both of them nearly twice as tall.

As the climate warmed, plants began to establish themselves on the hill, eventually carpeting it in the holy trinity of Puget Sound trees: Douglas fir, western red cedar, and western hemlock. Although we have no record, old-growth trees would have been taller than the height of the hill, making the hill look far bigger than it actually was, like a kid with an unruly mop of hair.

For thousands of years, the hill rose above the bay, periodically buffeted by storms and fires. The Native people who inhabited the region used the hill as a defensive lookout and must also have tickled its surface with hunting forays and perhaps beat down a trail or two. But they left little evidence on the land to tell of the specifics. Not until the 1850s, when the first European settlers arrived, did the hill change substantially. On February 15, 1852, William Bell and Arthur Denny filed papers with the federal government to claim ownership of the land. Bell and his wife, Sarah, claimed their property on the north, while Arthur and Mary Denny staked their acres to the south.

FIG. 5.1. DETAIL OF ELI S. GLOVER'S *BIRD'S-EYE VIEW OF THE CITY OF SEATTLE*, 1878. Library of Congress, Geography and Map Division.

Within a decade, the hill acquired its first name, when Arthur Denny set aside ten acres in 1860 at the southwest corner as a site for the state capitol. Although fellow pioneer Daniel Bagley convinced Denny that the state university would be a better choice than state government, and that Denny's property was "too far out" from the center of town, the name Capitol Hill clung to the high knoll.[6]

Our first accurate, or at least mostly accurate, view of the hill comes from E. S. Glover's 1878 *Bird's-Eye View of the City of Seattle*. Trees cover most of the summit and east side. Only a handful of streets, some of which barely stretch more than a single block, have been laid out, and most of them are on the northern, flatter portion of the hill, or what was becoming known as Belltown. Fewer than fifty houses dot these nascent blocks, with just a dozen or so on the steeper sections, including a curious octagon house. Built in 1875 by foundry owner John Nation, the three-story building stands on a bluff overlooking the corner of Front and Virginia Streets (the names *Virginia* and *Lenora* should be switched on Glover's map). The hill was so thinly populated that Front Street was the easternmost street for several blocks. A city directory shows that

FIG. 5.2. DETAIL OF HENRY WELLGE'S *BIRD'S EYE VIEW OF THE CITY OF SEATTLE, WT*, 1884. Library of Congress, Geography and Map Division.

most of the hill's residents worked with wood, as loggers, wood choppers, and carpenters.

By the time of the 1884 publication of Henry Wellge's *Bird's Eye View of the City of Seattle, WT*, most of the trees on the hill were gone. This was also the year that the beautiful Denny School opened on Battery Street between Fifth and Sixth Avenues. Two stories high and "handsome in the extreme," it was built on "Thoroughly Scientific Principles," attracting the children of the mostly working-class families that had begun to inhabit the hill.[7] These hill pioneers reflected the trades necessary to feed the city's constant growth. Brickmakers, stonecutters, blacksmiths, carpenters, painters, paperhangers, and plasterers all mingled in boardinghouses, duplexes, and one- and two-story homes. But foundryman John Nation had moved on, having sold the octagon house—what some called the "bird cage"[8]—to Orlando Root, a physician who had arrived in Seattle in 1878.

What had been a sparsely populated, recently logged open space had grown into a neighborhood. The Methodists and Episcopalians had opened their church doors, and a horse-drawn trolley skirted the hill on

Front Street. There was even a park, the city's only one, at the north end of the hill, plus a smattering of small shops.[9] The hill, though, had not yet earned its famous moniker. That would come in 1889.

The year before, Arthur Denny and several investors had decided to build a massive hotel on the hill's southern high point. As an early newspaperman, Thomas Prosch, wrote, "It was thought that if a large, showy, modern house were built upon an eligible, commanding site, with spacious grounds and grand view, properly managed and with the money-making idea of secondary consideration, that tourists from all parts of the country would be attracted to it, and that the town would be greatly benefited thereby."[10] With its 350 rooms and its price tag of more than two hundred thousand dollars, the Denny Hotel would be one of the grandest buildings in the city and one worthy of lending its name to the hill it crowned. Unfortunately for Denny and his partners, the Panic of 1893 led to financial difficulties that left an abandoned, unfinished shell, instead of a grand hotel, on the newly dubbed Denny Hill.

By the early 1890s, the hill had become part of the fabric of the city. Denny School had been expanded and now had twenty schoolrooms. Several trolley lines cut across the shoulders of the hill, providing access to jobs across the city. For many, Denny Hill offered a close-to-downtown place to live, separate from the more expensive homes on First Hill and the newly developing Queen Anne Hill. Others, though, saw something else. To Arthur Denny's granddaughter Sophie Frye Bass, Denny was the "Rock of Gibraltar . . . the highest hill in all the world."[11] The hill was also a natural border, wrote Bass, keeping gangs who lived in Belltown from spending too much time harassing kids who lived downtown.

◆ ◆ ◆

So where was this Gibraltar? As with most neighborhoods, no one had a precise definition. It depends on whether you look at it bureaucratically or topographically. From a regulatory standpoint, Denny, or at least its regrades, conformed to street boundaries, roughly between First Avenue, Westlake, Pike, and Broad Streets. In total, Denny encompassed about sixty-five city blocks, or a little over two hundred acres.[12] In 1893, there were at least 320 single-family and sixty-nine multifamily homes, along with twenty-six stables, 218 sheds, and ten hotels within this area.[13]

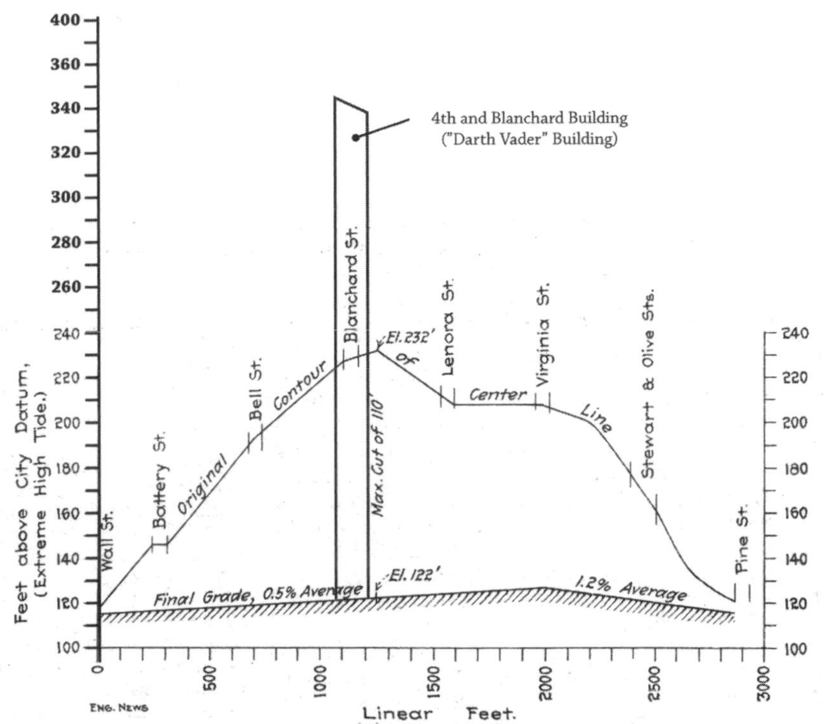

FIG. 5.3. PROFILE OF DENNY HILL ALONG FOURTH AVENUE. Compare the original grades (up to 12 percent) between Fourth and the northernmost of the hill's two summits. The original height of the hill (232 feet reflects a historic datum point) is about 100 feet lower than the height of the building that now stands at Fourth and Blanchard (aka the Darth Vader building), which is the location of the hill's former high point. This profile illustration comes from Jordan West Monez, based on George Holmes Moore, "Heavy Regrading by Means of Hydraulic Sluicing at Seattle, Wash.," *Engineering News* 63, no. 13 (March 31, 1910): 357.

Early topographic maps show the hill extending much farther north than most people probably think it did. Between the low point of the wedge at Thomas Street (across the street from the Experience Music Project and Science Fiction Museum) and its northern summit, at Fourth and Blanchard, Denny ran due south and gained 120 feet. The black wedges known informally as the Darth Vader building now stand at this spot. Topping out at a height of 344 feet from a base elevation of 135 feet above sea level, the apex of the taller wedge is about equal to what would have been the tops of the Douglas firs that once crowned the hill. In order to get a feel for the onetime elevation of Denny, I ventured up to the Darth Vader building's tenth floor, where a skeptical receptionist let me look out the windows. What struck me is that I could see Lake Union, to the northeast, something no longer possible from the modern street.

A second high point on the hill, the one where the Denny Hotel had been started, rose between Second and Third on Stewart Street. Gaining nearly a hundred feet from Pike Street to the base of the hotel construction, the hill's south side was a formidable obstacle, but it was hardly impassable. Trolley cars gained more elevation on their ascents up the city's other hills, and Denny's eastern slope was relatively gentle. In addition, the steep south face would provide the hotel with one of the most beautiful urban views in the country, a fact that Thomson reluctantly noted had attracted the many who moved to the hill.

He, in contrast, ignored the view, seeing only a "broken and irregular" mound that stifled commercial advancement. In a line memorialized by generations of writers, Thomson wrote that "Seattle was in a pit, that to get anywhere we would be compelled to climb out if we could." He resolved to "persevere to the end."[14] For Thomson and others, Denny Hill was the greatest impediment to going anywhere. "The crowning virtue of the regrades is to be found in the fact that it demonstrated that there could be found [a] sufficient amount of level land south of the Lake Washington Canal in one general vicinity on which to assemble the real center of the city."[15]

To Thomson's credit, he did not base his vision on personal profit. His biographer, Wilson, wrote of Thomson that "he altered Seattle's urban landscape . . . in order to improve the material conditions of living. Implicit in his work was a belief that an improved standard of living . . . encouraged enjoyment in life and spiritual fulfillment."[16] Throughout Thomson's writings he is almost religious in his constant concern that regrades would benefit other individuals as well as the city itself. Property on the city's hills, he wrote, "if left in its broken and inaccessible position in which nature left it[,] . . . would soon become of no commercial value whatsoever."[17] This was a stark contrast to the visions of people such as Semple and Burke, but then again, Thomson was a public servant, unlike them.

◆ ◆ ◆

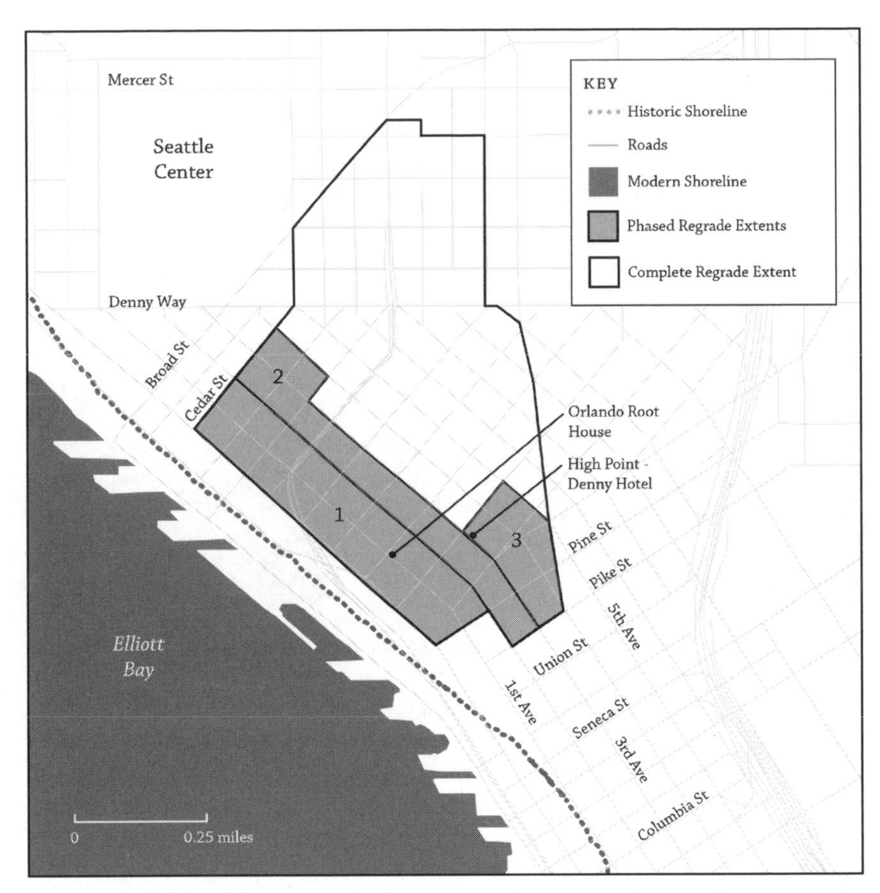

MAP 5.2. DENNY REGRADES 1 (1898), 2 (1903–6), AND 3 (1906–8).

After living in Seattle for more than fifteen years, Thomson, who had become the official city engineer in 1892, finally had a chance to focus on one of those arteries that blocked Seattle's future success.[18] In late 1897, less than six months after the beginning of the Klondike gold rush, the city council passed, with Thomson's guidance, an ordinance authorizing the regrading of First Avenue between Pike and Denny Way.[19] Although it was the smallest of all the regrades done on Denny Hill, the work on First Avenue set the stage for the many regrades to follow. Contractors used water from hydraulic hoses to wash away the hill. Most of the 110,700 cubic yards of sediment ended up in Elliott Bay, unused for anyone's benefit. At least one person delayed the process by obtaining an injunction against the city. The regrade, noted the *Seattle P-I*, was "endurable only

FIG. 5.4. WASHINGTON HOTEL, 1905. Taken during the regrading of Second Avenue, the photograph shows that Denny Hill and its famous hotel stood high above the city. Note the logs on the lower right side of the image, which were used to support the regraded hillside. Within three years of when this image was shot, the hotel and the hill under it had been removed. Courtesy Seattle Municipal Archives, image 77282.

for the sake of the promise of ultimate improvement."[20] Essentially, it was a typical Seattle transportation project.

With First Avenue washed into the bay, the city's shapers and scrapers could turn their focus to Second Avenue. Those who lived on Second seemed less than enthusiastic about lopping off their block, despite a significant rise in property values along First. When proponents began to organize in 1901 to gather signatures on the petition necessary for starting the regrade process, the residents of Second Avenue needed sixty-seven signatures, or 50 percent of those whose land would be affected. (Gathering the requisite signatures from property owners was the key step in the process.) After two years of discussions, enough owners signed, which triggered the next phase, asking the city council to pass an ordinance mandating the work. The ordinance passed on March 2, 1903, but the contractor did not start work until August. Because of the

FIG. 5.5. WASHINGTON HOTEL TROLLEY, 1903. James Moore built this electric trolley to carry passengers up Third Avenue from Pine Street to the hotel's entrance at Third and Stewart. The route was so steep that it required a counterbalance to assist the trolley in its block-long, hundred-foot ascent. Without the trolley, guests would have been "confined to the use of a hack, or a tedious, laborious climb," noted the *Seattle Times*. Courtesy University of Washington Special Collections, UW 4696.

geography and location of buildings, workers supplemented the hydraulic hoses with steam shovels to remove more than six hundred thousand cubic yards of material. Most of this sediment ended up deep in Elliott Bay, carried out by train over a trestle built at Battery Street.

The most notable event of the Second Avenue regrade occurred on August 19, 1905. In an act of sabotage, someone dug a deep hole across Virginia Street, effectively cutting off access to the southern summit of the hill, where Arthur Denny's great hotel stood. After sitting empty for a decade, the hotel had been purchased in early 1903 by developer James Moore, who renamed it the Washington Hotel. The son of a wealthy builder and shipowner in Nova Scotia, Moore had arrived in Seattle around 1886, quickly becoming a leading developer in areas such as the University District, Green Lake, and Capitol Hill. In his papers found at his death was a quote on urban planning by architect Daniel Burnham:

"Make no little plans; they have no magic to stir men's blood. . . . Make big plans; aim high and hope and work."[21]

No plans stirred the blood of Seattleites more than the completion of the Washington Hotel, which finally lived up to the vision of its original owners. Moore had fashioned one of the most ornate and beautiful hotels on the West Coast. To reach the hotel on the hill, he had built, in one week's time, a tramway up Denny's steep south face. Known as the "shortest streetcar line in the world," the stubby one-car trolley attracted visitors and residents who reveled in "dangling up Denny Hill," as locals referred to the ride.[22] The hotel's first guest was President Theodore Roosevelt, who stayed one night, on May 23, 1903.

Unlike many members of the Denny Hill community, Moore did not support the idea of regrading, or at least of regrading his property, and he filed several injunctions to stop the work on Second Avenue. Moore's impetus was that he wanted the lowest possible slope up Virginia Street to his hotel. In the days when horse-drawn wagons still ruled the road, a steep hill drove up transportation costs. According to a drawing found in the engineering archives of the City of Seattle, the cost per load could range from $0.50 to $2.50, with as many as nine horses necessary for 15 percent grades.[23] The original grade on Virginia between Second and Third was 14.8 percent. Grades on Denny Hill topped out at 22.65 percent.[24]

Moore suspected that I. E. Moses, a lawyer and real estate man, might be responsible for cutting off access to his hotel, in part because Moses was irate that Moore's injunctions had delayed the regrading of Virginia Street, which would have steepened the slope of the street from Second up to Moore's hotel.[25] The police never caught whoever dug the hole, but within a month Moore had worked out a deal to lift the injunctions and the regrading work continued.

This was not Moore's only battle over regrades. Two months after the completion of the Second Avenue regrade on March 4, 1906, a grand party was held in the Washington Hotel. The attendees had not come to celebrate the regrade's completion but to honor the hotel, which would soon be torn down. Months of negotiations between Thomson and Moore had led to what became known as the "Moore agreement," whereby Moore agreed to demolish the hotel at his expense, cede property for the extension north of Third Avenue, and regrade the hill down to the new roadway.[26] All he asked was that the city award him one dollar and pay for his assessments.

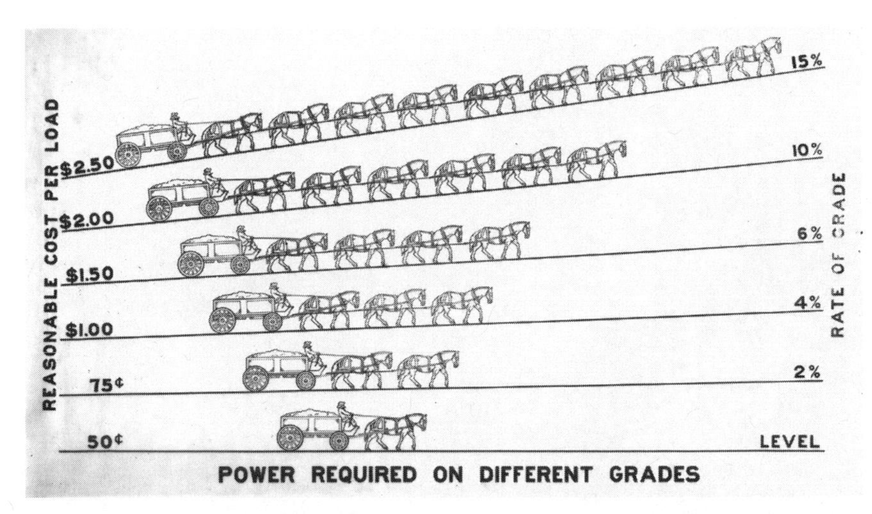

FIG. 5.6. HORSEPOWER TO PULL LOADS, CIRCA 1907 (THOUGH DATED 1913). In 1911, Reginald Thomson wrote in reference to this image, which had been printed on a card: "This card was carried about and shown to people in the district by a gentleman who had become intensely zealous for the regrade, and proved to be the most effective argument in favor of the regrade which any one could have given." Courtesy Seattle Municipal Archives, image 82.

Moore did not do this out of the kindness of his heart. "The great benefit being conferred upon the abutting owners" had so impressed Moore, that "he came before the City Council and asked leave to excavate all those parts of streets he had opposed regrading, and those which interconnected, and to conform to all the grades which I had laid down," crowed Thomson in his autobiography.[27] We have no record of Moore's specific justification for razing the Washington, but it seems that Thomson may have exaggerated Moore's change of heart. We do know, however, that Moore did rather well financially after cutting down the hill and hotel.

As historian Matt Klingle has written, Moore was unusual in that he had the "resources and connections" to fight the regrading projects.[28] He also differed from Burke, Semple, and Thomson in that there is little evidence that Moore hid his desires behind the Seattle Spirit. He set out to develop land with the simple goal of profit for himself.

One remnant of Moore's clash with the city remains. Stand at the corner of Second and Virginia and you will notice something odd—each

FIG. 5.7. WASHING AWAY DENNY HILL AT SECOND AVENUE AND VIRGINIA STREET, AUGUST 20, 1907. Razing Denny was a complicated project with trains, steam shovels, hydraulic cannons, thousands of feet of piping, and tons of debris. Note the pile of wood below the water cannons, which appears to be the remains of a house that fell into the pit formed by the regrading project. This panorama was made with two images taken by Asahel Curtis. Courtesy University of Washington Special Collections, UW 36359.

of the streets slopes down from you. No other intersection in the area of the former Denny Hill does this. Rising to 167 feet above sea level, the intersection of Second and Virginia is now the peak of the hill, or what one early writer called a "sort of terrestrial dunce-cap."[29] It exists because Moore's legal finagling prevented any additional dirt from being removed (about eleven feet had been lopped off) despite Thomson's goal of lower-

ing Virginia. Moore's hump was one of the few battles over regrading that Thomson lost; he was not happy about it.

Work on destroying the Washington Hotel and excavating the land under it began soon after the closing party of May 1906. Contractor H. W. Hawley won the contract and, working with steam shovels and hydraulic hoses, began to take away 650,000 cubic yards of the hill. Unlike in the other regrades, more than half the dirt that Hawley removed went someplace other than Elliott Bay.[30] A small steam railway colloquially known as the "Mount Moses, Denny Hill, Central and Westlake Valley Airline" ferried the dirt across a trestle and dumped it in the area around Pine and Olive Streets, creating the smoothed out, relatively gentle slope that now ascends past the Paramount Theatre to Capitol Hill.[31]

At the same time that his hotel was being destroyed, Moore was building a grand performance hall. The Moore Theatre opened in 1907, followed a year later by the New Washington Hotel, a fourteen-story building that would rival the old one for elegance. What makes these buildings interesting to modern residents is that they help illustrate what we lost with Denny Hill. Look to the top of the Moore Theatre; it is about equal to the hill's former elevation of 240 feet. Down the block, at the Josephinum, the modern incarnation of Moore's hotel, you can get a feel for how the old Washington Hotel must have towered over the city; the top of the Josephinum is about three hundred feet above sea level, roughly equal to the top of the former six-floor hotel.

Wanting to discover a sense of what it was like to ascend Denny Hill, I went into the Josephinum, which for many decades has been low-income housing, and got permission to walk up the fourteen flights of stairs. I started in the lobby, climbing up a wide stairway clad, like the rest of the lobby, in ornate black-and-white Alaskan marble. At floor three, a long corridor filled with the aromas of cooking food took me to another marble stairwell, where every eighteen stairs I ascended to another level. By floor five, I was breathing a bit harder. At floor ten I could look north to Lake Union, and by the top my pulse had quickened. Looking around, I could see many taller, newer buildings blocking my view. People on the top floor of the old Washington would not have had this problem. It was the highest building in downtown at that time, with spectacular views across, up, and down Puget Sound.

If you want a feel for the historic elevation of the north end of Denny Hill, look at the trees in Denny Park, which, before being regraded, was sixty feet higher than at present. The tallest, at about a hundred feet, are the tulip trees (the ones with the large four-lobed leaves) in the center of the park and the giant redwoods on the east side. Along John Street are eighty- to ninety-foot London planes (these have their species's characteristic patches of smooth bark flaking off the trunks), and in the center stand forty-five-foot false cypresses. Thus the park's original ground level was higher than many of the shorter trees in the park, though significantly lower than the tallest trees.

◆ ◆ ◆

As epic as the decapitation of Denny was, it was not the lone large-scale regrade in Seattle. Thomson also looked south and saw that access to the growing community in Rainier Valley was blocked by yet another "offensive" mound, Jackson Street hill with its 15 percent grades.[32] He was not alone in seeing these grades as a problem. In early 1904, members of the Rainier Heights Improvement Club requested that the city excavate two tunnels through the ridge made by Jackson Street, thus providing the ten thousand residents of Rainier Valley a better approach to the city.[33]

When Thomson heard about the idea for tunnels, he "gave the matter serious attention" but concluded that it "would be far cheaper and far better" to raze the entire hill—from Washington Street south for five blocks, to Lane Street, and along a narrow extension farther south for another seven blocks, to Atlantic Street. Twelfth Avenue would be the eastern extent of the regrade, which would cover fifty-six blocks, of which twenty-nine would be cut and twenty-seven filled. In terms of surface area, the Jackson Street project was the largest single regrade in Seattle.

For those who lived in the core of the affected area, "every house and every garden and every street" would be lost, wrote Thomson. But in his vision for the city, the "bare land remaining after such sacrifices . . . would be of more value than the property was worth as it then stood, and would be in a position to continuously increase in value as the city increased in population."[34] If nothing were done, added Thomson, the land values would drop to nothing. It's not clear why the values would plummet, considering that residents would still have been able to catch the trolley car that ran up Jackson Street every ten minutes and had been operating since 1888.

The contracting firm Lewis and Wiley started work at Eighth Avenue South and Lane Street in May 1907.[35] In order to wash away Jackson Street and the surrounding territory, they would need 6 million to 10 million gallons of water a day from the city system, 6 million more gallons per day from the city's abandoned Lake Washington pump, and up to 12 million gallons from Elliott Bay. To get the salt water, Lewis and Wiley had to encase four massive pumps in a watertight structure built on pilings two hundred feet out in the bay.[36] From there, a stave pipe made of Douglas fir ran for six thousand feet—under streets and railroad yards and over railroad tracks—to the hill. In order to keep teredos from destroying the pipe, it had to be flushed regularly with freshwater.[37]

The pipes from the various sources fed into the two pits where the work was being done—where the men operated the hydraulic hoses, or giants, as they were called, to blast away the soil. Each pit had two giants and a five-man team. One man fired each giant, holding a three-foot-long handle connected to a cast-iron nozzle. The seven- to ten-foot-long nozzle sat on an articulated base and had an adjustable opening from three to five inches in diameter. This was the most highly skilled position on the team and was often filled by "old giant [men]" who had learned their trade in Alaska's goldfields.[38]

In addition, one man on each giant team had a child's dream job: using boards to direct the muddy slurry that collected in the pit to a pipe, which carried the former hill out via a flume to the Duwamish tideflats. The boardman also removed roots or similar materials that could choke the system. Another man monitored the iron bars, or grizzly, at the mouth of the pipe, which kept out larger rocks and lumps of clay. He would either remove them or sledgehammer the larger material into smaller bits and let them go down the pipe.[39] Under normal conditions, a crew could wash away about a thousand cubic yards of material during each eight-hour shift, or enough sand and gravel to fill 6,750 wheelbarrows.[40]

We do not know who brought the giants to Seattle, but it is not purely coincidental that hydraulic mining technology, used extensively in the Klondike gold rush, appeared in Seattle shortly afterward.[41] The technology had been developed in 1853 near Nevada City, California, when a miner attached a wooden nozzle to a rawhide hose and blasted gold-rich gravel into sluices along a small creek. This use of high-pressure hoses quickly spread throughout the Sierra Nevada, leading to wholesale devastation as ever more powerful giants stripped hillside after hillside. Widespread hydraulic mining throughout the western United States may have contributed to why Seattleites so readily adopted this technique. Many residents, particularly those who had been involved in the Klondike, had perhaps seen the giants in the field or at least knew about them, so it would not have seemed that strange to use them in an urban setting.

Not coincidentally, Portland, too, used hydraulic giants to reshape its topography. In 1909, the men responsible for regrading Jackson Street, Charles Wiley and William Lewis, purchased part of Goldsmith Hill and sculpted it into a series of terraces for a neighborhood. The project was such a "novel idea," a reporter for the *Sunday Oregonian* wrote, that engineers from across the globe came to Portland to study it.[42]

Despite the reporter's enthusiasm, washing away one's hills was not easy, and no other city in the United States used this technology to alter

FIG. 5.8. HYDRAULIC GIANT, JUNE 6, 1906. Water shot out of hydraulic giants was the primary means of regrading Seattle's hills until the final regrading of Denny in 1929. Although the giants were effective, they regularly had to be supplemented by dynamite to remove hard sediments. The high pressure was achieved either by gravity feed, sending the water downhill through sequentially smaller pipes, or via pumps. Despite the power of the hydraulic giants, they required only one person to operate. Courtesy University of Washington Special Collections, UW 36352.

its topography. Hydraulic giants could not cut through solid rock; they worked best in sediments, such as river gravels in Portland and Seattle's glacial deposits. Denny Hill might still be here if it had risen on the hard rocks of the Puget Group and not on layers of poorly consolidated, glacier-derived material. Water was the other limiting factor. The giants required millions of gallons of water per day, which was relatively easy to acquire in Seattle. In contrast, when engineers proposed using hydraulic giants on Bunker Hill in downtown Los Angeles, it would have required building a twenty-mile pipeline from the Pacific Ocean. That plan never came to fruition.

◆ ◆ ◆

Watching hydraulic giants at work quickly became one of Seattle's major attractions; during the Alaska-Yukon-Pacific Exposition of 1909, many thought the regrades rivaled the great fair for spectacle. As a reporter for the *P-I* wrote, "You can practically tell how long a man has been in Seattle by the degree of interest he displays in these picturesque regrade operations." The thunderous roar of the water, the Niagara Falls–like spray, and the river of gray mud carrying away the despised hills kept a steady stream of visitors standing on the edges of the great pits. There was also the chance that an abandoned house would tumble or a landslide would cascade into the void. Though we might sneer at "Mr. Newcomer," concluded the *P-I* man, "come to think of it, we haven't seen it at close range for a few weeks and it's still interesting, isn't it?"[43]

Of all the engineering projects in Seattle, those that employed the hydraulic giants are the ones I would have most liked to have seen. I completely understand the appeal. It must have been hypnotic, like watching waves break on a shoreline but with the possibility of destruction and the periodic explosions of dynamite.

But regrade rubberneckers had to be careful, for scoundrels mined the crowd. When Everett florist William J. Bickert arrived in town, he, like many newcomers, found his way to the pits, where he struck up conversation with a couple of men. After watching the water with his new friends, who soon offered an excuse and slipped away, Bickert discovered that they had pilfered his wallet filled with fifty-five dollars in gold. Nor were the pits always safe for workers: youthful hooligans regularly tossed brickbats and rocks down toward the men and equipment, causing injuries and damage. (The pits were dangerous places, and death struck in many fashions. An explosion caused by an employee thawing twenty sticks of dynamite in a pan over an open flame killed a nine-year-old boy. One worker drowned in a flume that carried material out to Elliott Bay. Another died when he dropped his nozzle and it struck him. A third got caught between the massive belts used to carry away the dirt on a later regrade, and several were buried by landslides or falling rock.)

Lewis and Wiley completed the Jackson Street regrade in December 1909.[44] They had moved 3.35 million cubic yards of dirt while cutting down the hill by eighty-five feet at Ninth and Jackson and while adding about thirty feet at Sixth and Weller.[45] Not content with leveling Jackson Street,

FIG. 5.9. WATCHING THE REGRADE, JULY 1907. Few things were more intriguing in 1907 Seattle than its regrades. At least forty people, including several on a porch, are looking into the pit of what is probably the project that removed the Washington Hotel. To our modern eyes, it's amazing how little protection separated viewers from the steep drop into the pit. Courtesy Seattle Public Library, image from Seattle Regrade Photos, R709.7972 W399S.

Thomson also cut down the hill at Dearborn Street. The maximum cut was 106 feet, at Twelfth Avenue, now spanned by the bridge that connects Beacon Hill to the city.[46] The material from Jackson and Dearborn went to create eighty-five acres of land on the east side of the tideflats.[47]

The Dearborn cut is the one regrade that Thomson seems to have considered less than completely successful. In contrast to the results of other projects, the "greater portion of its length proved an injury to the immediately abutting property rather than a benefit," he wrote.[48] Thomson justified the project, however, by stating that although there was no local benefit, the community benefited from the improved connection between distant parts of the city.

◆ ◆ ◆

Thomson had not forgotten about Denny during his rush to cut down Jackson. By holding public meetings and going door to door, he and his assistants had acquired the necessary signatures for a petition to regrade the untouched parts of Denny. This time they went for the hill's heart with a plan to chop off the entire west side, from Fifth Avenue to Third Avenue between Pike and Denny. They turned the petition in on April 28, 1906. Mayor Moore (no relation to James) signed Ordinance No. 13776 on May 23, 1906. It established what property the project would remove. A second ordinance, number 14993, provided the funding mechanism, or what was known as a Local Improvement District.

First used in Seattle in 1893, the Local Improvement District made the landowners pay for the improvement of their own property, as opposed to forcing the city as a whole to pay. How much a property owner owed was based on a simple formula. On the plus side was the appraised value before regrading: the owner would be paid for any buildings destroyed during the project. The owner would also add to the plus column any theoretical damages to the property caused by lowering the street. On the negative side were the costs of rebuilding the infrastructure, which were divided among all who benefited, in proportion to how much their property would benefit. In order to determine these values, a judge appointed a commission, in this case three "young and ambitious" men—two lawyers and an engineer.[49] They held a series of public meetings, visited the properties, and sent out inquiry letters, after which they presented their findings. If the owners didn't agree with the commission, they could appeal through the courts.

Of course, many perceived the process as unfair, which in some ways it was, though *arbitrary* might be a better term. The commission made mistakes and probably was influenced by politics, but many of those poor decisions were corrected by appeal, and only a handful of cases made it all the way to the state supreme court. Thomson's biographer, William Wilson, who has written the most detailed account of the Denny Hill

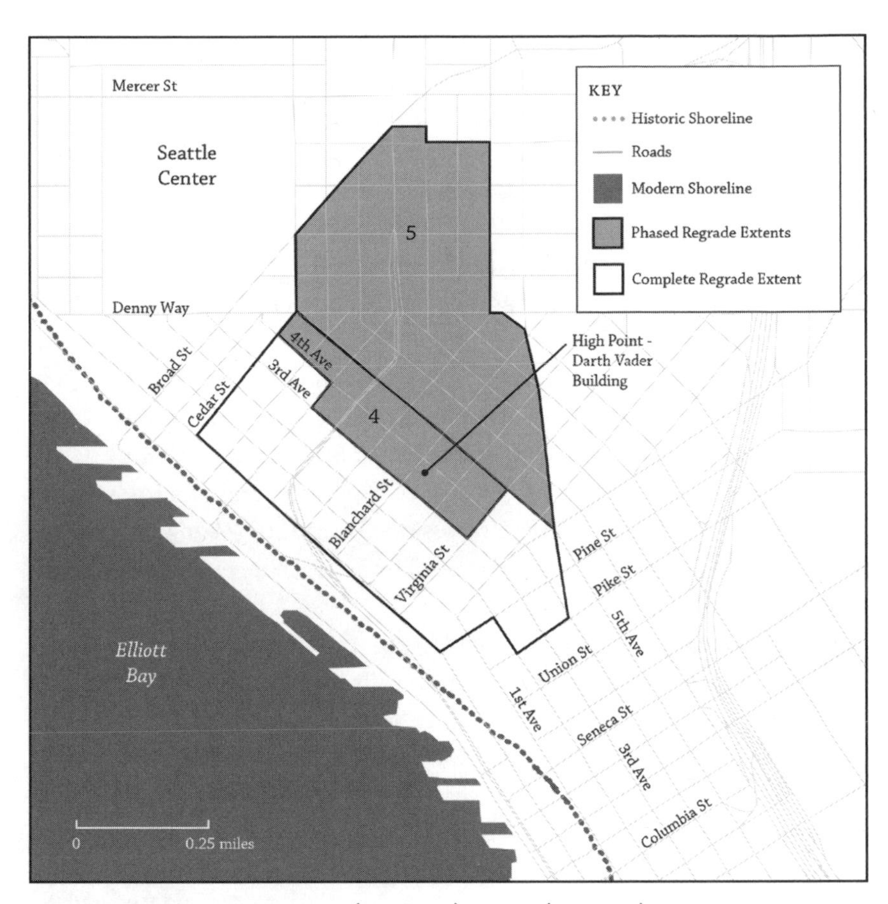

MAP 5.3. DENNY REGRADES 4 (1908–11) AND 5 (1929–30).

Local Improvement District and appeals, told me, "The problem didn't so much [lie] with the process as it did with what happened afterward."[50] Homeowners had ten years to pay back their assessments, at a rather steep 6 percent interest. In addition, they had to pay taxes on the new, generally higher assessed value of the land. They also had to pay the contractor separately for lowering their property. Many could not afford to make these payments.

With most of the court cases settled, the city called for bids to be opened in July 1907. Rainier Development Corporation won the contract for the Denny Hill regrading project with a bid of $0.27 per cubic yard to regrade the city streets. In signing the contract, Rainier also agreed to excavate all private property at the same price that the city paid, but only if landowners agreed to have their property excavated at the same time as the city property. One year later, Thomson sent a letter to Rainier instructing them to "begin work forthwith."[51] They would have thirty months to complete the removal of 5.4 million cubic yards of Denny Hill. This long-desired regrading of Denny Hill began when subcontractor Grant Smith and Co. and Stilwell started washing away Third and Bell in August 1908.

The hill that GS&C&S attacked with hydraulic giants in 1908 was not the hill of 1884. With Seattle's population having grown to a smidge over two hundred thousand, little undeveloped land existed in or near downtown. Denny was no exception. Buildings covered the hill, of which more than 450 were residential structures. Single-family houses dominated, but there was also a mixture of apartments and other multifamily dwellings, as well as shops, three churches, two schools, and one fire station. There were also at least twenty stables on the hill, reflecting how people still relied on horses for transportation. By 1930, there would be only one stable left.

The demographics had changed, too. By the early 1900s, Denny attracted a more diverse community than it had in the 1880s, though it was still almost entirely white. As an example, consider the high point of the hill, Fourth and Blanchard.[52] At the northwest corner of what was still a dirt road, in a two-story duplex with four units, lived Walter Camp, secretary and treasurer of the Pioneer Carriage Company, and his wife; a widow, Adelaide Poore, who rented rooms to a painter and a wagonmaker; James Smith, an insurance agent, and his family; and Frank Rogers and his wife and son. Across Blanchard was former city councilman Thomas E. Jones, who had settled on the hill in 1884, when he was listed as a farmer; he was now a pile driver who had worked on many projects in the tideflats. He shared the house with his wife, his mother, and five children, ages six to eighteen, the younger of whom walked three blocks to the beautiful Denny School. A biographical description of Jones noted that he lived in a "comfortable, modern home, in which much taste and good judgment are displayed in plain though comfortable furnishings."[53] The house had two floors and a prominent front porch.

Dillis B. Ward lived next to the Joneses. A schoolteacher and real estate speculator who had arrived in Seattle in 1859, Ward had made two fortunes and lost both, the latter in the Panic of 1893, which forced him and his wife, daughter, and son-in-law to downsize to Denny Hill, or what Ward nicknamed "Poverty Row."[54] Around the corner on Fourth

FIG. 5.10. ANSON BURWELL HOUSE. Few high-quality shots exist to show the elegance of the homes that formerly stood on Denny Hill. Anson Burwell's exemplifies the style of houses built in the 1890s. Note that the street is not paved but is dirt. Courtesy John Goff.

was a large house, formerly the home of John Megrath, a contractor who had built the United States Post Office, customs house, and court building. He had moved out of that house in 1905, and the Young Women's Home Club now occupied Megrath's old house. It had been established as a safe, inexpensive home for young ladies, many of whom worked for low wages in the department stores downtown.

Another ex-councilman lived down the block from the Wards and Joneses. William V. Rinehart lived with his family and two boarders. The Rineharts had purchased their property in October 1885 from William Bell's widow, Sarah, as had the Burwell family, which bought two side-by-side lots at auction for $915. Brothers Austin and Anson Burwell had built twin three-story homes next door to the Rineharts. Austin, the elder brother, and president of the Seattle Cracker and Candy Company, lived with his extended family, including his sister and father. Next door was Anson and his family. Like many people on the hill, Anson worked downtown—as manager of Seattle Hardware Company—and probably paid a nickel daily to catch one of the trolley lines that ran on the west side but not over the hill. The brothers would sell their "expensive and beautiful" houses together for $25,000 in December 1905. A real estate article in the *Seattle Times* noted that "nothing will be done in the way of further improvement until the regrading of that district is finished."[55]

Despite the influx of people and businesses on Denny, nearly everyone, or at least those whose voices have made it to the present, felt the way Thomson did. The hill blocked Seattle's future, stifled progress, and prevented the unhampered flow of business into its natural channels. The hill might be pretty and picturesque, but for any city hoping to become world class the hill was a great handicap and needed to be removed.

◆ ◆ ◆

Before the contractors could begin their assault on Denny, they had to determine where to get the water and what to do with it after the giants had converted the hill to mud. Unlike the Jackson Street regrade, this one would not use any city water. Fortunately, GS&C&S could suck water out of Elliott Bay using the pumps built by H. W. Hawley for his work on Denny Hill under the Washington Hotel. The Hawley machines provided 5 million gallons a day. Supplementing their saltwater source, GS&C&S installed three pumps on Lake Union and fed 18 million gallons a day through a mile-long pipe up Eighth Avenue to two distribution points on Battery Street, at Third and Fifth Avenues.

As a comparison, the entire population of Seattle consumed about 23 million gallons of water a day, about a million gallons less than was being used on the concurrent Jackson Street regrade.[56] This meant that during every twenty-four-hour period, the two regrades used about twice as much water as the city. It was only a decade earlier that Seattle had begun to tap the Cedar River for its drinking water, and ten years before that the city hadn't had enough water capacity to put out its Great Fire. (One of the general items of common misinformation about the regrades is the use of city water. On Denny Hill none was used, and only about one-quarter of the Jackson Street water came from the Cedar River.)

To get rid of Denny, it was decided to dump the hill into the deeper water of Elliott Bay. The muck would travel via an 875-foot tunnel excavated from Bell Street and Second Avenue to Elliott Avenue.[57] The timber-lined tunnel, tall enough for a man to stand up in, led to a flume, or sluice, that passed over the tracks and roadway of Railroad Avenue and discharged the hill into the water. By the time the contractors finished regrading the hill, the flume extended on pilings for 1,200 feet into Elliott Bay. It was an engineering marvel, since they had to use 125-foot piles, which they drove into previously dumped fill, in water that had been 200 feet deep before the regrades began. In order to spread the fill widely, the flume had three distribution points, which allowed the pile driver to extend one arm of the flume into deeper water while the other two continued to dump Denny.[58] Because of all the sediments, the turbidity was so bad near the dumping site that fishermen complained it had driven cod away from the docks and out to clearer water.

Throughout most of this phase of the project, water provided enough force to dislodge the hill; but workers had to rely on dynamite to supplement the giants when they hit hard lenses in the blue-gray clay found in the lower parts of the hill. They also needed to blast when they hit hard layers of glacial till or hardpan, and when the workers came across boulders. More exciting was when workers found fossils, including tree trunks and mammoth teeth, several of which the Burke Museum now owns.

Work began slowly, with workers excavating only 8,000 cubic yards of material in August 1908. It didn't help that technical difficulties meant that the contractor got only four days of work done that month. The total jumped to 42,000 cubic yards in September and 130,000 cubic yards in November, with a peak of 294,500 cubic yards in October 1909. By November 1910, more than 97 percent of the hill had been removed.

◆ ◆ ◆

What remained was, in part, what has become one of the most famous aspects of Denny Hill and Seattle history: the isolated buttes known as spite mounds, spite heaps, or spite humps. In May 1910, six of these notorious pillars of private property rose above the flattened surface of the hill, giving the site the appearance of a surrealist's mash-up: desert Southwest meets urban sprawl. Five property owners had not paid to have their land lowered, supposedly protesting against the city's plans to level the hill.

As far as we know, however, none of the owners were against cutting down Denny. One of them, Zachariah Holden, had owned his house at Fifth and Virginia since 1883, when it was "far beyond the hum of the

FIG. 5.11. SPITE MOUNDS, 1910. Perhaps the most famous, or infamous, photograph of the Denny regrades, this image is often described as showing spite mounds. (It is not unusual to see it printed in reverse.) The house in the upper left was in the process of being moved from Fifth Avenue one block east to Fourth Avenue. James Kelley owned the two mounds on the right, the closer of which stood at Fourth and Bell; the farther one stood at Fourth and Virginia. Photo by Asahel Curtis, Courtesy University of Washington Special Collections, UW 4812.

traffic of the little town." He told a *P-I* reporter that he hadn't signed up to have his property lowered, because he didn't have the money needed to pay for the work. Holden eventually was able to borrow the necessary funding to pay to eliminate his "spite mound." "I want to see every vestige of these hills come down," he said. "I would have been willing to make greater sacrifices to see Seattle become greater still."[59]

James T. Kelley owned two of the mounds, one at Fourth and Virginia and one at Fourth and Bell. Each rose about a hundred feet. A former owner of a small meat market, Kelley had gone north in search of Klondike gold and made a fortune. He was in Alaska working his mines when the contractor had sought authorization to cut down his property. When he arrived back in Seattle in the summer of 1910, Kelley, too, said that he supported the removal of Denny. "I have lived too long in Seattle . . . to be accused of being actuated by spite on any matter connected with public improvements. After full consideration of the matter and in the belief that the Denny Hill regrade district should be rapidly improved, I have decided to improve my holdings."[60] By January 1911, both mounds were gone.

Judging from Kelley and Holden, I am not sure that spite is part of the story of these buttes. The *Oxford English Dictionary* defines spite as a "strong feeling of hatred or ill-will." It doesn't seem to me that a lack of money and an absence from home meet that definition. Kelley and Holden, like most of their neighbors, heartily supported the Denny regrades. They appeared to do so for two reasons. Many must have agreed with Thomson, that the change would benefit the entire city, and that an improved Seattle with easier transportation routes, more space for industry, and room for businesses to spread would be a better place to live. Others also believed that they would make money by being able to sell their lowered property.

Surprisingly, there is little additional information in the local papers about the mounds beyond a *P-I* article from January 10, 1910, which names three of the owners but gives no other details. City records show that John Megrath, the man who owned but didn't live in the Young Women's Home Club, objected to his assessments and filed protests with the city. Yet he had also signed the initial petition requesting the city to regrade the area under his property.

Nor is there evidence for another common urban belief, that homes were left on the mounds, forcing the owners to reach their residences by ladders, steep stairs, or mountaineering equipment. There are photographs showing isolated homes on what appear to be buttes, but no one wrote about or photographed anyone commuting vertically to them.

The most famous of these hovering homes is a stunning three-story Victorian with six chimneys, numerous porches, multiple dormers, columns, and oriel windows. Situated on the southeast corner of Fourth and Lenora, the rooming house had at least ten apartments available for rent, most of them furnished. It had been built in 1903 by

FIG. 5.12A. MICHELSEN HOUSE, NOVEMBER 1909. In the foreground are several people working to control the water and sediments unleashed by the hydraulic giants. Courtesy Museum of History and Industry, image SHS15452.

FIG. 5.12B. MICHELSEN HOUSE, DECEMBER 1909. Courtesy Museum of History and Industry, image 1998.55.19.

FIG. 5.12C. MICHELSEN HOUSE, JUNE 1910. Most of the regrading was finished when James Lee took this photograph. Note how the Michelsen house sits off the ground, waiting for a new first floor to be built under it. The house to the left and in the background is the large house in the famous Curtis photograph (fig. 5.11). The two mounds are the ones owned by James Kelley. Courtesy University of Washington Special Collections, Lee 20024.

FIG. 5.12D. MICHELSEN HOUSE, 1912. This photograph is one of the best illustrations of the flat, barren nature of regraded Denny Hill. The Michelsen house has now been shorn of most of its Victorian decorations. The solitary house across the street from it is the house peeking out from behind the mound in the previous photo. It is a unique home in that it is the only house I could find that was lowered to the spot directly below its pre-regrade location. Courtesy University of Washington Special Collections, UW 5026.

Mark Michelsen for ten thousand dollars. A photo taken in November 1909 shows it perched on the top of a bluff, high above three hydraulic giants. Only part of the house is on the ground; the south side rests on cribbing, suspending the house anywhere from five to ten feet off the ground. One month later the entire house is shown airborne, supported by stacks of cribbing, the highest of which contains at least thirty-five cribs, or pallets. Next, in June 1910, we find the house in the middle of the regrade, about ten feet above the flattened earth just east of the unleveled butte owned by James Kelley. Finally, in 1912, the house rests on terra firma, though on a new first floor that had been inserted under the old structure. Now located two lots south of its original site, the building has lost its elaborate gables and chimneys and is known as the Michelsen Apartments. It remained at that site until new owners leveled it in 1956.

Homeowners on Denny Hill had several options with their homes. Some simply abandoned them after paying to have their property regraded. These owners probably hoped to sell their lowered property at a good profit. Such homes might be torn down before excavation or allowed to fall into the pit during excavation, where they could be burned, hauled off to the refuse heap, or scavenged. Many owners did as Michelsen did and moved their home to a new location. According to the minutes of the Board of Public Works, which granted permission to move homes, nearly every week between 1908 and 1910 at least one or two owners applied to move homes. Some moved them as much as a mile.

One who did was John Megrath. In April 1909, he applied to the board for an application to move his old house, and a smaller one he owned on Third Avenue, about three-quarters of a mile to Harrison Street and Fifth Avenue North, due east of the modern-day Memorial Stadium at the Seattle Center. We don't have records of how Megrath's house was moved, but he probably worked with Lemuel B. Gullett, who regularly advertised in the *Seattle Times*: "No job too large to stagger me. None too small to receive attention." Gullett appears to have been the main mover in town, or at least the one who took out the most ads in the papers. Not only would he move houses for those who wanted to continue living in them, he also had clients who would buy houses on Denny Hill and pay to move them to safer locales.

Gullett's first step with the Megrath house would have been to cut or drill holes through the foundation directly under the sills and insert twelve-by-twelve-inch timbers perpendicular to the direction of travel. Spaced about three feet apart, the timbers would support the house and provide a surface for raising it. The house probably contained all its furniture, even fragile items such as picture frames and mirrors. Raising the house was done with jackscrews, a device similar to a modern car jack, which required a crew coordinated with whistles in order to turn the screws simultaneously on the many jacks necessary to raise the house. During the moving of the three-story, brick Eagles' Hall in 1909, the mover used 480 jackscrews; a small house might require only four jacks, one at each corner. Gullett then lowered the house onto wood carriages, typically three feet long and eighteen inches wide, with iron spikes and wheels. The carriages sat on rails made of wood or iron. Running out from the house were ropes, attached either to oxen or horses or to a windlass-and-pulley system, which could be cranked to pull the house forward. As the house progressed, the rails would be picked up and brought around in front of the house.[61]

After Megrath's house arrived at its new location, Gullett would have reversed the process, using jackscrews to lower the house to its new foundation. Sometime later, the ladies of the Young Women's Home Club followed—apparently Megrath moved the home so that the women could keep living in it—and paid three dollars a week for room and board.

The Department of Public Works declared the regrade to be complete on June 20, 1911. In April, steam shovels working at the corner of Sixth and Virginia had finished the final trimming of the rough edges left by the giants. By the time the work was complete, more than 5.5 million cubic yards of the hill had been dumped into Elliott Bay. Everyone who had lived around Fourth and Blanchard was gone. The Rineharts' house was the only other house moved off the hill, though the Rineharts did not follow it.[62] They moved up to Capitol Hill, as did the Megraths, Camps, Anson Burwell, and Ms. Poore. The Wards and Austin Burwell's family migrated up to Queen Anne, as had the Joneses. None of the Jones children still attended Denny School, but if they had, they would have found it a far different building. (The engineers had taken their job seriously. Since part of the great building had extended across the official border of the regrade—the alley between Fifth and Sixth Avenues—the west wing had been chopped off and the school now perched on a forty-foot-high cliff.) Only Frank Rogers stayed nearby, moving less than a mile to downtown.

◆ ◆ ◆

Now that half the hill was gone, businesses were supposed to flow into the new and natural channels. But before any could, Thomson and others came up with a new set of plans for the hill. In a way these were even more ambitious, and vastly more expensive. If removing Denny was a topographic manifestation of Thomson's vision to better connect the city, then the new scheme would carry out his ideals via a grand transportation plan.

Following Thomson's advice, a city committee hired Virgil Bogue in September 1910 to come to Seattle and develop a comprehensive plan for civic improvement in Seattle. Bogue was best known in Seattle for bringing the Northern Pacific Railway through Stampede Pass, and for a failed 1894 plan for developing the Duwamish tideflats. Working with a crew of engineers and surveyors, Bogue was able to present his plan to the city planning commission in August 1911. After swift approval, Bogue's Plan of Seattle went to voters in March 1912.

It is clear why Thomson advocated hiring Bogue. Bogue might have been one of the few who had even greater disdain than Thomson for the city's topography. In addition to desiring to regrade Denny Hill again, Bogue proposed to bypass Seattle's hills, "which in other vicinities would be dignified as mountains," via a series of tunnels.[63] One would go under Lake Washington, another under Beacon Hill, and a third under the north end of Capitol Hill. The combined length of the tunnels would be five and a half miles.

In addition, the report's central focus was what Bogue calls the Civic Idea, or "preparation for a probable population of the future." He defined it as conscious recognition that "community life is not an aggregation of unrelated parts and functions, and agglomeration; but a growth, a product, whose many elements are virtually interwoven and inter-dependent." To achieve this Civic Idea, "correct planning for a city requires . . . the ideal arrangement of main arteries[,] . . . which will enable traffic to flow from any point in the city to any other point in the city."[64] Thomson couldn't have written it better.

Much of the rhetoric used by Bogue came out of the City Beautiful movement, which had started to spread after the World's Columbian Exposition, or Chicago World's Fair of 1893. At the movement's core, city planners sought to improve their cities and their citizens through grand architecture and great public spaces. The beautiful structures and pleasant parks and roads would revitalize people, as well as inspire social, civic, and moral virtues and perhaps provide economic benefits. Or as Thomson wrote, "It is impossible to determine in measurable units, the financial, hygienic or aesthetic benefits that have resulted from these works [the regrades]."[65]

Although Seattle's topography generally precluded Bogue's planned road system, he had found a point in the city where he could direct a series of arteries to bring people into Seattle from the north, south, and east. That point was Fourth and Blanchard, where he proposed to build a civic center worthy of the world-class city that Seattle would become. Oval-shaped, the center would encompass six grand municipal buildings, including a city hall and courthouse, library, and art museum.

Part of what made Fourth and Blanchard so appealing to Bogue was its recent regrading and corresponding lack of buildings. This meant that the city would not have to purchase buildings through condemnation. And since commerce had yet to advance this far north, property would be cheap, perhaps as little as $3.5 million for the entire fourteen-block civic center. In addition, Bogue noted without any apparent irony, the new, low elevation meant that the great civic center could be seen from all of the surrounding hills.

It was truly a miracle. Everything that Thomson said would happen would. Denny Hill had evolved from a despised, ungainly piece of landscape to become the very center of Seattle, the point at which all roads would intersect. Not only would it be the business center but it also would be the center of a thriving civic life, a place of grand ideas, of legislation, and of law.

There were just a few problems. The Civic Plans Investigation Committee, a group composed primarily of landowners from around the old business district, estimated that Bogue's plan would cost more than $100 million. They also pointed out that the proposed civic center would require raising the ground for the buildings and streets between fourteen and thirty feet. After the city had spent nearly $1.5 million and forced hundreds of families to relocate, Bogue now proposed to rebuild part of the hill. As one letter writer to the *Seattle Times* put it, "No sane man would ask for such a scheme."[66] He was right. On March 5, 1912, Seattle's citizens soundly rejected Bogue and Thomson's vision for Seattle: 24,966 against, 14,506 for.

◆ ◆ ◆

Nor apparently did business owners recognize the merits of Thomson's opening up the land north of the historic downtown core. Seven years after the last regrading project, little had changed on the former Denny Hill. This is evident in one of the few maps that show Denny between the regrading projects, Seattle's Coming Retail and Apartment-House District, published by architect B. Dudley Stuart. James Moore's towering New Washington Hotel and Moore Theatre anchor the southwest corner of the old hill. Running along the east side is an abrupt bluff, topped at its north and higher end by the delimbed Denny School. Toward the north was a smattering of low buildings, including a livery, lumberyard, and fuel yard. And at Fourth and Blanchard, the hill was now a flat plain surrounded by four blocks with only one house on the land. The one thriving business opportunity in the regrade area was film exchanges, hardly the wave of businesses expected by regrade proponents.[67]

The situation on the remaining part of the hill (between Fifth Avenue and Westlake Avenue and Virginia Street and John Street) was more dire. A *Times* article called the area "Seattle's Old Quarter" and referred to "its unpaved streets, its unpainted weather beaten houses, its absence of traffic, its general air of somnolence." By 1920, the number of single-family homes on the hill had dropped to 178, from 361 in 1900. Few owners had done any upgrades on their houses because they knew that civic progress might strike the hill before the paint had time to dry. But for children, the Old Quarter was a paradise where they could play marbles or skip rope in the middle of streets, watching out only for the rare flivver or delivery cart, reported the *Times* writer.[68]

Many of those who did live on what remained of the hill were transient. The average child attending Denny School had lived at six addresses in his or her short life, with many having resided in more than ten homes. More than half of the students at the start of the school year at Denny were gone by the end of the year.[69] These children were the sons and daughters of camp and mill workers, many of whom would take their families back to work camps in early spring. One advantage to living on Denny was the availability of furnished apartment and rooming houses, which allowed families to live out of their suitcases and skip out when they couldn't pay their bills.[70]

Planning for eradication of the Old Quarter started in 1923. By this point, planners had moved beyond the horse as a means of transportation; the automobile was ascendant and central to how Seattle would plan for the future. There is little mention of the earlier justification for regrades—allowing businesses to move north of downtown. The regrade was now about getting people past the hill and into the downtown business district.

In contrast to previous regrades, the contractor did not use hydraulic giants. Too much of the area surrounding the project had been developed, so funneling a muddy stream across paved streets heavy with traffic was not practical. Previous efforts had also shown that the giants always needed assistance from dynamite and power shovels, or excavators. The new regrade's excavated sediment, though, ended up in Elliott Bay, carried to the water's edge by an extensive conveyor belt system, and transported out into the water by one of the engineering marvels of the Denny regrading projects: self-dumping wooden scows.

Designed by naval architect William C. Nickum, these self-dumping scows were mirror image top and bottom, with open decks that held four hundred cubic yards of dirt. Between the decks were two internal tanks, one on each side of the scow. A tug towed the full scow out into Elliott Bay, where a crew member pulled a rope that opened valves, or seacocks, on one side of the boat. Within three minutes, water filled one of the tanks, and the out-of-balance scow flipped over, dumping its load. No longer weighted down by the dirt, the scow rose high enough to drain the internal tank, which took about eight minutes. The tug then pulled the scow back to shore, ready for its next load.

All this equipment had to be built or acquired before excavation could begin. The hill also had to be cleared of buildings, including two city landmarks. First to go was Denny School, destroyed in late 1928. Before the great building could be completely leveled, a hastily formed alumni association gathered enough money to save the forty-eight-foot-tall cedar cupola. It weighed nearly sixteen tons. Placed in Denny Park, it remained a source of pride for thousands of Denny School alumni until 1950, when park employees burned it to the ground because of its state of disrepair.[71] One block south stood the stone-and-brick Sacred Heart Church. It did not succumb easily. The bell tower withstood more than twenty sticks of dynamite, falling only under a final, epic assault. Many watching the attempts felt that the tower's tenacity was "a divine rejoinder against man's reducing process."[72] A similar fate of destruction

FIG. 5.13A. SELF-DUMPING SCOW FULL OF SEDIMENT, MAY 8, 1930. Scows would be loaded with material carried by conveyor belt to a dock that extended out into Elliott Bay. Courtesy Seattle Municipal Archives, image 4076.

FIG. 5.13B. SELF-DUMPING SCOW BEING TOWED OUT INTO ELLIOTT BAY, MAY 8, 1930. Barges towed the scows out into the bay. Courtesy Seattle Municipal Archives, image 4077.

FIG. 5.13C. SELF-DUMPING SCOW BEGINNING TO TIP, MAY 8, 1930. After an underwater valve was opened, water would fill a chamber on one side of the scow, causing it to tilt. Courtesy Seattle Municipal Archives, image 4078.

FIG. 5.13D. SELF-DUMPING SCOW EMPTYING ITS LOAD, MAY 8, 1930. As the chamber filled, causing disequilibrium, the scow would begin to flip and dump its load. Courtesy Seattle Municipal Archives, image 4080.

FIG. 5.14. DENNY SCHOOL, 1904. Built in 1884 at the northeast corner of Fifth Avenue and Battery Street, Denny School was expanded seven years later with two four-room additions on either end. In 1910, workers cut off the west wing of the building and left it perched forty feet above Fifth Avenue. Finally, in late 1928, the building was razed as part of the final regrade of Denny Hill. Courtesy Seattle Public Library, image spl_shp22683.

befell the rest of the Old Quarter's more than two hundred buildings, including several that had been moved from the hill during the previous regrade.[73]

Preliminary work on cutting down the last of Denny Hill began in February 1929, with a power shovel removing dirt along Fifth Avenue for construction of the main conveyor belt. By May, the contractor was ready to go at Denny with two big electric shovels on Battery Street at Fifth and at Seventh. With all five shovels in operation, each day would see the hill reduced by more than twelve thousand cubic yards.

By December 1930, the hill was almost gone. The crews had been unrelenting. After a slow start in 1929, the contractors had worked out their

FIG. 5.15. DENNY SCHOOL CUPOLA, 1929. Built from cedar logs cut at Yesler's mill, the cupola was saved from destruction at the last minute by alumni of Denny School. The plan was to preserve it for posterity, noted the *Seattle Times*, although by 1936 it was in poor enough shape that the Parks Department threatened to destroy it. Finally, in 1950, it did. Courtesy Seattle Public Library, image spl_shp22684.

problems and were completing a task that many thought impossible just months earlier. Bite after bite of the hill was fed into the hoppers, dropped onto the ever-moving conveyor belts, and shot along at six hundred feet per minute to the waiting scows and the dirt's final voyage out into Elliott Bay.

On December 9, 1930, shovel operators spent the day racing each other to "clear 'er up" for a final celebration. "That once great promontory, thrown up by nature ages ago to make trouble for Seattle's progressive expansion," was now merely one shovelful of earth, noted the *Times*. "At last the way was opened for giant new buildings, for the forward sweep of the central business district northward on level grades into territory

FIG. 5.16. FIRST SHOVELFUL ON DENNY HILL, MAY 11, 1929. In May 1929, an electric shovel began work on the final regrade of Denny Hill at Fifth Avenue and Battery Street. Each bucket dumped two cubic yards of material through a funnel hopper onto a moveable conveyor belt, which carried the dirt to the main conveyor belt and ultimately to the scows that ferried the dirt out into Elliott Bay. Courtesy Seattle Municipal Archives, image 3383.

which only a few months ago barred progress with a lofty barricade of clay banks, ill-kept streets, run-down residences, aged public and semi-public structures."[74]

At 11 A.M. the next day, Mayor Frank Edwards climbed into one of the electric shovels between Battery and Wall Streets, near Sixth Avenue, and pulled a lever, and the final cubic yards of the hill vanished. Hundreds stood by, including many who had been born on the hill, who had attended Denny School, and who had lived there. "There were many moist eyes," wrote a reporter for the *P-I.*[75] With Edwards's final dumping of the dirt, 4,354,625 cubic yards of the hill had been removed. Total cost was $2,261,800.

FIG. 5.17. LAST SHOVELFUL ON DENNY HILL, DECEMBER 10, 1930. A gleeful Mayor Edwards pulls a lever to remove the final shovelful of Denny Hill. Courtesy Seattle Municipal Archives, image 4613.

The hill was gone. Almost. Two holdouts still resisted having their mounds cut down to street level.

At the north end stood the Taylor Apartments. Built in late 1905, the building had eight units. But it would soon be gone, since the owners had given in and signed a contract to have the building destroyed and the land lowered.

One block over, at 307 Sixth Avenue North, between John and Thomas, the home of Frank C. Brownfield sat atop a small knoll. Born in Seattle in 1886, he was the grandson of Christian Brownfield, who with his wife had, in 1867, homesteaded the land that would become the University District.[76] Five years later, Christian, along with Frank's father, Curtis, was hired to pull coal cars across the land connecting Lake Union and Lake Washington. Once the cars were across, they went onto the steamer *Clara*, which Curtis piloted down the lake to meet Seattle's first railroad.

Curtis moved to Denny Hill in 1886, when his wife, Harriet, gave birth to Frank, who was most likely helped into the world by their neighbor, the owner of the octagon house, Dr. Orlando Root. When he was old enough, Frank walked the couple blocks from his family's house on Fourth Avenue over to Denny School.

By 1903, Frank, his brothers James and Clyde, and his mother and father were living at 307 Sixth. According to Frank's son, Curt, the two-story house didn't have indoor plumbing because Frank's father didn't think a toilet should be in a house. Instead there were two toilets on the back porch. The family washed their clothes on the opposite side of the porch.

"My parents were not very cooperative. They didn't see why they should pay anything extra," Curt told me in March 2014. "They just wanted to live there. They weren't going to benefit from the regrade." So they decided not to go along with everyone else and instead chose to leave their house perched atop a mound of dirt. Nor did they pay any taxes or assessment, which ultimately led to the family renting the house back from the city for ten dollars a month.

"When they took away all the houses, it was nice for us kids, as we had all this real estate to run around in," said Curt. They also benefited from the area becoming a sort of dumping ground for heaps of metal scrap and tires. Curt and his friends would rummage through the junk piles, finding useful items such as artillery and even an Enfield rifle. On October 21, 1942, his mother was quoted in the *Times*, saying, "In normal times these junkpiles would look terrible. Now that we are at war, however, and know that all this scrap will be used against the enemy, it's really a beautiful sight."

To get up to the house, which was about twenty feet above the surrounding terrain, they had a wide ramp cut into the hill. It led to a flight of stairs up to the level ground of the house, where Curt's father had built a chicken coop. In front, they could climb a steep metal staircase, which came from an old ship that had been scrapped out. The family used the back entrance because it was closer to the trolley tracks and because the front staircase had been removed.

"It was a little different," says Curt. The house was somewhat of a point of interest, but he added that, during the Depression, everyone was struggling, so few people made a big deal out of it. By the early 1940s, Curt's older brother had made enough money to buy a new house for

FIG. 5.18. BROWNFIELD HOUSE, 1937. Standing all alone on its mound, Frank Brownfield's house was the final house on Denny Hill. Before the regrade, a flight of steps led up to this doorway, which was the front door to the house. This image was part of a Works Progress Administration project that photographed every property in the county. Courtesy Puget Sound Regional Branch of the Washington State Archives from the King County Assessor's Office Property Record Cards.

his parents, and the family moved to the top of Queen Anne Hill. "My brother and sister were a bit embarrassed by the house." Moreover, his sister was always worried the house would slide off its perch.

A ship worker then moved into the home. He didn't stay long, though, because on August 11, 1945, real estate operator James L. Napier, who had purchased the property, filed for a permit to destroy the house. Later, Seattle City Light built a power substation that still stands on the site of the former Brownfield house.

"People all around us were sold on the idea that once this area was flat, the property was going to become very valuable and developers were going to come in," said Curt. "It was the wrong presumption. It never happened. The Depression took over and business went downhill. Most people lost everything. We didn't lose everything because we didn't go along with the regrades. It wasn't a bad decision. It was a good decision not to have our house moved."

With the razing of the Brownfield house and mound, the hill was finally no more. No one held a ceremony. I suspect that no one even commented on Denny's demise; with a world war going on, there were far larger concerns in Seattle to think about. But the leveling of the little house on the hill was an important symbol, for after five decades of discussions, disputes, disruptions, and destruction, the long-hoped-for elimination of Denny Hill was complete. The sixty-five blocks that had constituted a towering hill with two summits and steep sides had been transformed into a planed-off landscape, flat enough for any means of transport.

One can argue that its passing had been noted at the ceremony in December 1930, but Brownfield's holdout had for more than thirteen years been a provocation, an irritant to those who felt that nature had no right to stand in the way of progress. The house and the mound were a tangible reminder that, once upon a time, a great hill had stood at the north end of the downtown business district.

◆ ◆ ◆

Amazingly, R. H. Thomson was still alive when Frank Brownfield's house and mound finally succumbed. He was eighty-nine years old and working on his memoir, which would come out in 1950. We have no idea if he even knew of the Brownfield home and its persistence, but he was in Seattle on the hill's fateful last day in 1930. In his journal he recorded: "Last work by steam shovel . . . on Denny Hill regrade." At the time, Thomson was the city engineer again. He had been rehired in August 1930.

Thomson had not been city engineer since 1911, when he quit to start work for the nascent Port of Seattle. After a short stint at the port, he had spent most of the intervening years consulting on engineering projects throughout the Northwest. Mayor Edwards had called Thomson out of retirement following the untimely death of the previous engineer. Because the final regrading of Denny was so far along, Thomson had little to do with it, though there is little doubt that he supported the project. As late as 1929, he was still offering plans for cutting down hills in Seattle. In his memoir he wrote dismissively of those opposed to regrades: "Some people seemed to think that because there were hills in Seattle originally, some of them ought to be left there, no difference how injurious a heavy grade over a hill may be to the property beyond the hill."[77]

Thomson's attitude certainly reflects the thinking of his generation. Most of the contemporaneous histories of Seattle declared regrades to have been necessary, as well as a grand success. They also dismissed the "public that could not grasp the boldness of his [Thomson's] conception nor appreciate the comprehensiveness of the work he was doing."[78] Curiously, one of the few to question the regrade of Denny Hill was Thomas Burke. He told a reporter at the closing party for the Washington Hotel that "from a commercial point of view, and certainly from an esthetic one, it would have been much better to have saved Denny hill by carrying Third Avenue under it, thus obtaining the desired result, while preserving in all aspects the natural beauty that means so much to any city."[79]

Although I tend to dislike Burke and his arrogance, I agree with him about Denny Hill. I see little reason why it had to be removed. Photographs show that it was a lovely residential complement to downtown, with numerous grand buildings. Nor had it stifled any development; the city continued to expand north throughout the early 1900s and into the teens. In addition, trolley service provided access to the hill, and cars were just beginning to appear, which would have rendered any objection to the steep slopes moot.

Access up the hill was certainly not a problem by the time of the final regrade, when cars had become widespread. City planners also should have realized that businesses were not migrating north into the new landscape made by earlier cuts as quickly as some had hoped they would. It seems that the final work occurred more out of inertia than any other reason. With most of the hill gone, it was inevitable that the remaining high points had to come down. It certainly didn't help that Denny Hill's population had become poorer and had little ability to prevent the destruction of their neighborhood.

As Thomson's biographer observes, attitudes about whether cutting the hill down was right or wrong began to change in the 1970s, particularly with Roger Sale's *Seattle: Past to Present*, an opinionated and entertaining history of the city. "The regrading of Denny Hill turned out to be little benefit and may well have been positively harmful," wrote Sale. He felt, much as I do, that the city suffered when it lost the Washington Hotel and a fine neighborhood that could have ranked with those in San Francisco. (If you want a feel for what was lost, look at the south end of Queen Anne Hill and its remaining turn-of-the-century homes.) Even worse, in Sale's view, was what had happened in the years after the work was completed. The neighborhood had become a "chronically semiblighted area," home to car lots, garish motels, and cheap furniture stores.[80]

Although historians such as Sale have looked at this period and the lack of any significant commerce flowing into the area as a sign of failure, it is an irony that the buildings from this not-so-grand era of commerce are some of the loveliest and most valued at present. Many have been torn down, replaced by glassy, modern high-rises, but those that have survived are the buildings that have earned historic designations, that have been converted to restaurants, cafes, offices, and condos, that give Belltown its hip and trendy feel. They are the buildings that have helped lay the groundwork that enabled the former Denny Hill to fulfill its long-believed-in destiny: its transformation into an essential part of downtown Seattle, complete with large buildings, thriving businesses, and good lines of connectivity.

◆ ◆ ◆

The modern builders and developers share the same optimism that enticed those living on Denny Hill to regrade their neighborhood. They truly believe that they inhabit a growing and striving city with nothing ahead but a rosy future in which they stand to make a good profit. Ultimately, I think that optimism is what united the land-alteration projects in Seattle. Many people who had come west to live in Seattle in the late 1890s and early 1900s had come because of the growing economy, in particular after the establishment of transcontinental train routes and the boom of the Klondike gold rush. They expected, or at least hoped, to do well financially out here. If that meant carting off the ground below their homes, then that's what they would do, again optimistically hoping that they would benefit.

Another reason for the willingness to level our hills was that Seattle had an unusually mobile population. Around 20 percent of men eighteen and older who had arrived in 1900 were still here in 1910, a rate significantly lower than the rate in other American cities.[81] Combine this with a young city, where families didn't have generations of history, and razing your home probably didn't seem as daunting as it would be on the East Coast, where one family could have resided in the same house for longer than Seattle had been a city.

Finally, Seattleites possessed a dose of insecurity. Whether they were comparing Seattle to Tacoma, Portland, San Francisco, or New York, Seattleites have long sought to prove their city the equal of others. This is an ideal recipe for reshaping the landscape; all we have to do is fill in this spot, raze that one, or drain a third and we will achieve our destiny of greatness.

Why shouldn't the earlier generations of Seattleites have held these beliefs? In the eyes of people like Arthur Denny, Eugene Semple, Thomas Burke, and particularly Reginald Thomson, every landscape change improved the city, making it easier to get to and easier to move through. There was more land for industry, more land for transcontinental trains, more land for commerce. Reshaping the topography had changed Seattle from an isolated gathering of hopeful settlers, who had in many ways not chosen the best spot to establish a town, to undisputedly the premier city of the state.

By the time the last dirt was removed and Denny Hill was no more, landscape manipulation had become part of the city's collective DNA. It is how we looked at our world, how we viewed the future. Or as Sale wrote, "One of Thomson's most enduring legacies has been the habit of mind which thinks that all problems of any city can be solved by more and better engineering."[82] We may be proud that Seattle is not as flat as Kansas, extol the views created by the topography, and revel in the beauty of the landscape, but deep down we agree with the words of real estate agent Von Tarbill: "The Hill was too high and too steep to be utilized for business purposes, so that inevitably . . . the Hill had to be removed, somehow or other."[83]

FIFTH AVE COURT.
SAMSON
TIRES
PARKING
E & RICHFIELD

6

We Shape the Land and the Land Shapes Us

NOWHERE CAN ONE GET A BETTER FEEL FOR WHAT HAPPENED AND is happening to the Seattle landscape than the observation deck atop the Space Needle. Staring out to the regraded Denny Hill, the city, the sound, and the surrounding mountains, I find that what leaps out most, at least from a topographic standpoint, is the visible handiwork of our recent glacial past: this is a damned hilly place. Nearby rise Capitol Hill, Queen Anne, West Seattle, and Magnolia, with more hills stretching north, accentuated by the low, late afternoon light. To the east, the hills drop down to Lake Washington, and Bellevue, Redmond, and Kirkland—a band of faded ridges—stand serried before the mostly snowcapped Cascades. And most dominant of all: the stunning Mount Rainier, the towering ice-cream cone that seems far less than sixty miles distant.

Flat just doesn't seem to fit the vocabulary used to describe this region, which is why we both love this landscape and feel a need to change it. For many Seattleites, the scenic beauty of the unforgiving topography is what has long defined the city and made it unique and interesting and a desirable place to live. But flat places beckon, too, with their orderliness and adaptability.

FIG. 6.1 (OVERLEAF). DENNY HILL BEFORE (APRIL 16, 1928) AND AFTER (SEPTEMBER 22, 1931). The profound transformation of the final Denny regrade is evident in the lower photograph. What hasn't changed, though, is the Fifth Avenue Court apartments; the building still stands at the southeast corner of Fifth Avenue and Blanchard Street and even has the same name. On the left side of the upper image you can see the cupola of Denny School and the brick bell tower of Sacred Heart Church. Note the pathway leading up the slope to Denny School. Courtesy Seattle Municipal Archives, image 9331.

LYON VAN
& STORAGE CO
GENERAL GASOLINE
BEFORE
(Photo taken April 16, 1928)
DENNY HILL No 2 REGRADE
Contract Awarded Sept. 14 1928
Area 92 Acres - 38 City Blocks - Max. Cut 89 ft.
4,216,158 Cu. Yds. Earth moved by belt conveyor
to Elliott Bay and dumped by automatic scows.
Contract Under Supervision of City Eng.

AFTER
(Photo taken Sept. 22, 1931)
DENNY HILL No 2 REGRADE
Contract Completed Aug. 9, 1931
Total cost completed improvement including
123,100 Sq. Yds. Con. Pavt., 20,204 Lin. Ft. sewers
14,246 Lin. Ft. C.I. Watermains, 33,900 Sq. Yds. Con.
Walks and removal of earth from priv. prop. $1,885,240
Contract Under Supervision of City Eng.

I sense this dichotomy as I look below. We took down Denny Hill so we could mold the new landscape to our business needs; and yet, what is being built there now but tall buildings, as people try to recapture the missing elevation and beautiful views that the hill once afforded? I argued somewhat facetiously in my book *The Seattle Street-Smart Naturalist* that it looked as if developers were trying to assemble a steel-and-glass framework to support a new Denny Hill. That impulse doesn't appear to have slowed down in the decade since I wrote that book. From the Needle I count one white, two red, and six yellow cranes at work in the area of the Denny regrades.

I am further struck by how hard it is to pick out the Moore Theatre and Josephinum. Knowing that they mirror, respectively, the height of Denny Hill and the top of its most famous building. I realize how inconsequential the hill would seem in our present landscape, and that it really was not that big back in the day. Perhaps the problem of Denny Hill was that it wasn't big enough. If it had been as tall as Queen Anne or Capitol Hill, maybe no one would have proposed getting rid of it. Reginald Thomson and his ilk would have been too daunted, but Denny was just low enough that it didn't scare them.

The lack of stature of the old theater and hotel makes the hill's regrading seem ever more astounding to me. How could early Seattleites have considered it to be so horrible and in the way? Of course, that is my twenty-first-century mentality, compared with a late-nineteenth- and early-twentieth-century viewpoint. I find it hard to envision when horse-drawn carts and carriages were a principal means of travel, when one had to worry if it was physically possible to transport goods up and down steep slopes, or when Seattle really was way out west and days away from the East Coast by train.

Nor can I envision the mentality behind a project on the scale of the regrades, or of the similarly scaled cutting of the ship canal and filling in of the Duwamish tideflats. These were massive endeavors equal to any attempted by cities across the United States, and they were completed when Seattle was relatively small, at times more a town than a city. Perhaps most astonishing, two of the three projects were more or less driven by the desires of one man: Eugene Semple on the tideflats, and R. H. Thomson on the regrades. No one in Seattle today has the power, or chutzpah, to attempt something like these men did.

It helped that Semple and Thomson, as well as Burke and Chittenden, had few regulatory issues to slow them down. As far as I can tell, there was no consideration of the environmental consequences, or of the psychological consequences to the Native people and the people whose homes and businesses were on the regraded hills, in the way of the ship canal, or along the lakes or tideflats. If anyone attempted a similar project now, there would be committees, meetings, studies, reports, and probably a couple of votes, and then a new mayor or city council would get voted in and we'd start the debate all over again. But let's cut to the chase: in reality, none of the historic projects that I discuss in this book could ever take place in modern times. There are too many rules and regulations, too many people to object to them, and too much money involved.

◆ ◆ ◆

I don't think we can even comprehend how they did what they did. Consider one of the big projects started in 2010, the so-called fixing of the traffic corridor known as the Mercer Mess, just east of the Space Needle in the überhip South Lake Union neighborhood. We have demolished buildings, widened underpasses, and realigned several streets, but nothing on the order of removing a hill or making hundreds of acres of new land. And yet, when I look down from the Space Needle at the Mercer project, I count nine large excavators, compared with the five much smaller ones used on the final regrade of Denny Hill.

I can also see into a pit of Seattle's most famous modern megaproject: the State Route 99 tunnel. Within the block-long hole are towering cranes, bulldozers, excavators, concrete pumpers, and moving dots of yellow workers, all surrounded by thirty- to forty-foot-tall, stout wooden retaining walls. Located between Aurora and the headquarters of the Bill and Melinda Gates Foundation, the pit will be the exit point of Bertha and the deep-bore tunnel, which will replace the elevated roadway of the Alaskan Way Viaduct.

Although these huge projects cost billions of dollars, they are still relatively small compared with earlier ones, at least in regard to rearranging earth. Bertha is supposed to excavate a 1.7-mile tunnel, which may seem impressive, but it will move just 850,000 cubic yards of sediment. During the peak four months spent regrading Denny in the summer of 1909, the hydraulic giants and steam shovels moved 942,000 cubic yards of soil. And Bertha was stopped for more than fifteen months by a pipe, some plastic, and a boulder; in the old days, people would have laughed at so trivial an attempt at blockage, and then tossed a few sticks of dynamite, cleared out the debris, and carried on.

Despite their comparatively small size, each of these modern projects would have pleased our earlier generations of land surgeons. Both epitomize the long-term goal of improving access to the city's central business district. One of the justifications for the final razing of Denny Hill was a plan to extend Dexter, Eighth, and Ninth Avenues across the leveled landscape directly to Fifth Avenue and into downtown. When those new roads didn't get built, planners began to look elsewhere for access but lost momentum during the Depression and World War II. Planning did not pick up again until Seattle entered a postwar boom. Along with economic growth came an increase in the popularity of cars and the independence they provided, as well as the growth of suburbs, which meant that more and more people would need faster and easier access to and through Seattle's urban core.

A report published in 1947, and based on thousands of interviews, concluded that the best way to meet this need was to build two new roadways: one along the waterfront and one running along the west edge of Beacon Hill through the University District.[1] Because the waterfront route was cheaper, about $6.3 million, planners chose what would become the viaduct over the future Interstate 5.[2] Erecting a raised roadway along the waterfront was also a natural continuation of Seattle's long-term process of trying to elude its topography. After all, the viaduct would essentially be an elevated mirror of the original, in-the-water trestle of the Seattle Lake Shore and Eastern Railway, built in 1887. Both had the same goal, best expressed by Thomson as "enlarg[ing] insufficient arteries" and allowing the "free flow" of commerce and people.

Work on the double-decked viaduct began in February 1950. The concrete bypass opened on April 4, 1953, with the *P-I* calling it the "royal necklace across the bosom of the Queen City of the Pacific Northwest."[3] Others, too, likened it to a necklace, but one that was strangling the city and choking off access to the waterfront.

As the decades passed and the waterfront lost its utilitarian, train-choked, industrial feel, more and more people clamored to remove the viaduct and make the area where it ran more hospitable. Doing so, they said, would help make Seattle a world-class city by reconnecting downtown to the historic waterfront. The antiviaduct crowd failed to point out that taking down the concrete blight would not return the city to a more natural setting; the city would simply be eliminating a structure built in the 1950s so that modern citizens could access a completely artificial landscape, which had been finished only in 1936, the year the seawall was completed. In addition, few opponents of the viaduct advocated a return to the working waterfront that defined Seattle for most of the twentieth century.

A more compelling justification for removing the viaduct followed the 2001 Nisqually earthquake, which weakened, cracked, and settled the roadway. With the viaduct gone, Seattleites wouldn't have to worry about getting crushed by crumbling concrete, though it is unclear how safe we will be after we construct new buildings atop the liquefaction-susceptible fill behind the seawall.

In the years after the Nisqually quake, the viaduct removers gained momentum. Of course, Seattleites still dithered about how to keep a transportation corridor along the waterfront and even voted against a tunnel. Finally, in January 2009, state and local officials came together to support the deep-bore tunnel that is supposed to emerge from the pit below the Space Needle. The plan calls for a two-mile tunnel under the former Denny Hill. Excavated by Bertha, the four-lane, double-decker roadway will run from near CenturyLink Field to about the Space Needle, with its deepest section more than 270 feet below ground, right under Moore's high mound at Second and Virginia.[4]

Building a tunnel to create a better transportation route is hardly a new idea in Seattle. Since at least 1890, when real estate entrepreneur L. H. Griffith proposed to excavate a tunnel under Denny Hill to access his development in Fremont, engineers, planners, and developers have looked with lust at the opportunity to penetrate the city's hills to speed access from here to there. In most cases, the late-nineteenth- and early-twentieth-century proposals, and even some twenty-first-century ones, sound as if the tunnels were a foregone conclusion. Some of this has to do with hyperbole on the part of the planners and newspapers, but it still amazes me that so many people proposed so many tunnels, particularly at a time when the only way to build one was with shovels, picks, and wheelbarrows—though they did do rather well with those primitive tools. I cannot tell if it was naïveté, fantasy, or a stubborn disregard for topography that drove their plans. Whatever it was, most of the proposed transportation tunnels didn't make it past the hyperbole phase.

Despite not building these pie-in-the-sky (or pie-in-the-ground) projects, over one hundred tunnels snake for a total of more than fifty miles under Seattle. Most are much smaller than ones needed for transportation and are for moving sewage and, therefore, little noticed or known by the public. Since the mid-1980s, though, big transportation tunnels have become the modern equivalent of regrading: the most visible, and expensive, manifestation of the Seattle Spirit and our preferred method to subvert our topography.

The first large modern project brought Interstate 90 under the Mount Baker ridge into downtown, followed soon after by the Metro bus tunnels through downtown. Neither could have been built without the use of novel technology. And as happened when Joe Surber built the Seattle and Walla Walla Railroad and Transportation Company trestle across the Duwamish tideflats, the initial projects showed what was possible and prompted modern-day planners to go through hills more often than over them. By 2021, we will have Sound Transit light-rail tunnels cutting through Beacon Hill, downtown, and Capitol Hill; they will connect Capitol Hill with the University District, the University to the Roosevelt neighborhood, and Roosevelt to Northgate.

Bertha and its problems have attracted the most attention focused on tunnels, but the modern projects rely on the same technology, and on the fact that relatively soft, glacial sediments—and not hard rock, such as granite or sandstone—make up the hills. Tunnel-boring machines consist of a multitoothed, rotating cutting head that spins 0.1 to 2.5 revolutions per minute; a corkscrew-style muck conveyor that carries the spoils away from the cutter face; and a tunnel-liner assembly apparatus that places premade concrete panels into the structural rings that make up the tunnel. Once a ring is complete, jacks in the boring machine push off the ring to move the entire machine forward. Because of this design, the machines cannot go backward. Trailing behind the basic machine is support gear, which in the case of Bertha stretches for more than three hundred feet and includes the dirt-transport train, supplies, and restrooms. About twenty-five people work each shift.

The machines range in size from 21 feet in diameter for the light-rail tunnels to 57.5 feet in diameter for the State Route 99 tunnel. The main reason we hear so little about the non-Bertha machines, known as Brenda, Balto, and Togo, is that they have worked essentially as planned, excavating as much as a hundred feet a day. Perhaps size does matter, and smaller is better.

◆ ◆ ◆

Not only has Bertha sucked all the attention away from the other successful tunnel projects, but it also has drawn the focus away from the city's other billion-dollar waterfront project. From the Needle, I can look out along the entire length of the seawall, the concrete, wood, and steel barrier that protects Seattle from the sea. Built between 1916 and 1936, it has done its job well. But it is in need of replacement, and in January 2012 the citizens of Seattle voted in favor of a $290 million, thirty-year bond for a new seawall.

Running from Washington Street to Broad Street, the sixty-four-hundred-foot seawall is designed to protect the waterfront from storms and earthquakes, as well as from the favorite foe of the seawall's public relations team, gribbles. Along with shipworms, gribbles are responsible for ravaging the old seawall's wooden parts. Not that the designers are antiwildlife: the new wall also includes features that will enhance habitat for salmon, kelp, and waterfowl. The final environmental impact statement reports that the new seawall will "help preserve downtown Seattle and the region's economic competitiveness and quality of life."[5]

As part of improving our quality of life, and a direct outgrowth of razing the viaduct and rebuilding the seawall, urban planners are also at work on plans for a waterfront park. The twenty-acre public space is proposed to run from CenturyLink Field along the waterfront and into Belltown. With cantilevered vistas, trees galore, mist-shrouded boulders, thirty-six-foot-wide sidewalks, and outdoor cafes, the park would finally allow Seattleites to "reclaim the waterfront" and create "a front door to our city," or so say its proponents.[6] Opponents note that there would still be up to seven lanes of traffic along the waterfront, that every drawing shows the park on a beautiful sunny day, and that there is no mention of what private development will take place adjacent to the park or acknowledgment of either the waterfront's Native history or its industrial history. Total cost for the park as proposed would be several hundred million dollars.

The designs for the new seawall and park, and the removal of the viaduct, reflect not only new technology and new knowledge—little was known before the 1950s about how susceptible Seattle is to earthquakes—but also a new relationship to our surroundings. The rhetoric of decision makers now includes discussions of the importance of green spaces, of restoring ecological communities, and of developing recreational amenities for visitors and residents. Landscape manipulation may be in our DNA, but we also want such projects to benefit us psychologically and ecologically as well as economically. Our genetic code is adapting and evolving to changes in our relationship to Seattle's rolling topography.

This evolution is most visible in how we deal with our new understanding of how our physical environment can change and is changing. Where once we saw ourselves as the agents of change, now we worry we will be the victims of landscape change. For example, planners know that sometime soon, at least geologically soon, we will experience an earthquake with the potential to change Seattle forever. Buildings will tumble, hillsides will slide, roads will crack, and lifelines will break. The question is not *if* but *when*. We can see this specifically in the seawall, where potential earthquake damage was a major driving force in its design. If the seawall fails, Alaskan Way, the viaduct, and the buildings below it could all fail catastrophically. The new seawall is supposed to protect us from this fate. The new tunnel is also designed to withstand a magnitude 9.0 earthquake, which tunnel designers call a twenty-five-hundred-year quake.

More definite and predictable change will also come through rising sea levels brought about by our warming planet. A report published in 2008 predicts a sea-level rise in Puget Sound of three to twenty-two inches by 2050 and six to fifty inches by 2100. The models show that a rise of this magnitude will lead to flooding in many low-lying areas, such as Georgetown, Harbor Island, Interbay, and Ballard. What makes the numbers more of a concern, and Seattle more vulnerable, is what happens when sea-level rise combines with storm surges, a situation that we have already gained some insight into.

This occurred on December 17, 2012, during what has become known as a king tide.[7] These extreme high tides occur when the moon, earth, and sun are aligned, the moon is closest to the earth, and the earth is between the moon and sun, all at the same time. Compounding the problem of astronomically induced high tides that December day were a pair of weather phenomena. First was a low-pressure system over the region, which initially caused the sea level to rise. (The water rises because the atmosphere acts like a giant hand pressing down on the water. Lower pressure relaxes the hand, allowing water to bounce up.) The second phenomenon was the strong winds pushing onshore, which helped drive the water up onto the beaches. The resulting tide was the highest ever recorded in Seattle.

By 6:30 A.M., ninety minutes before high tide, waves were breaking high against the seawall at Alki Point. At high tide, water completely submerged the sandy beach; rose over the seawall, swamping the plaza where the mini Statue of Liberty stands; and flooded many houses around the point. The Duwamish River flooded the South Park neighborhood and backed up storm-water pipes, including one where water nearly shot out of a maintenance hole more than a mile from the river. Where the hole is located is adjacent to a low spot predicted to flood as sea level rises. (Coincidentally, it was about one block away from where the famous May Day picnic of 1874 was held.)

This surge could have been worse if we had had significant precipitation. The additional water would have raised the level of the Duwamish, which would have led to more over-bank flooding and put more water into storm pipes. Rainwater that would normally drain off city streets into storm pipes would not have been able to enter the already full pipes, resulting in another source of flooding.

"King tides are the leading edge of sea-level rise," says James Rufo-Hill, climate specialist for Seattle Public Utilities.[8] When you have one that combines with a low-pressure storm and high winds, it shows what will be more common in Seattle's future. Low-lying areas will be flooded, bluffs will be eroded, beaches will lose sand, the infrastructure will be stressed. Further compounding the effects of the rising sea level is that climate models predict greater frequency for storm events, which means an increase in the probability of storms and of king tides that will exacerbate their effects. Rufo-Hill says that extreme high tides won't happen daily, but tides higher than what we now experience will become a more regular occurrence.

◆ ◆ ◆

"There are three response strategies to address the predicted changes: retreat, accommodate, or protect," says Lara Whitely Binder, outreach and adaptation specialist with the University of Washington's Climate

Impacts Group and coauthor of the most authoritative analysis of sea-level rise in Washington State.[9] Retreat is the most unpalatable of the three; no one wants to admit that nature has the upper hand. To do so would involve political and economic issues that few officials want to address, which is why the ongoing rise in sea level has not resulted in any official retreat in Seattle. The city has, however, more or less retreated in some landslide-prone spots by creating regulations that would make rebuilding on the slopes of hills very expensive. In addition, King County spent $12.1 million to buy out and relocate a mobile home park that was situated along the Cedar River, in part because of an expected increase in frequency and intensity of flooding due to climate change.

Accommodation is the most common response. For example, Seattle Public Utilities has begun to raise the electric works at their substations to keep them above rising tides. King County has started retrofitting pump stations and storm pipes to prevent salt water from entering the storm-water system, and the Port of Seattle is considering different ways to reduce the effects of salt water on concrete docks. In each of these situations, the public agency recognizes that change is coming, and that they have to address it, looking not just at costs but also at who they serve, how they serve them, and what those services are.[10]

The most visible example of protection is the new seawall. It has specifically been designed to accommodate the sea level projected for 2100, as well as for storm surges at high tide. The designers calculate that the combination of a maximum estimated sea-level rise occurring at the same time as a fifty-year tide *and* a storm event is likely to happen once in a thousand years and would result in roughly three feet of water overtopping the wall. "This is an acceptable risk, given the fact that sea-level rise can be monitored over the life of the structure," says a member of the Seawall Project Street Team.[11]

No matter what labels we use, preventive action is how we now respond to our topography. We no longer focus on how to change the landscape because it's in the way of progress or because we need new areas for industry. Nor do we worry as much about the effects of topography on transportation. Because of our dependence on cars, we have insulated ourselves from the hills to the point that about the only time they concern us is during snowstorms.

◆ ◆ ◆

Ironically, the topography that concerns us most is flat and is largely the land we altered long ago. Look at the King County Flood Control District's Liquefaction Susceptibility map, and you'll see that the areas of highest concern are the former tideflats, Interbay, the Renton waterfront, and the Center for Urban Horticulture. On the City of Seattle's Sea Level Rise: Year 2050 map, the major trouble spots correspond directly with modified areas, primarily along the Duwamish River, including several spots that historically were shoreline and are now inland.

Both of these maps pinpoint Harbor Island as the area most susceptible to liquefaction and flooding. The island is what Ron Harmon, formerly the emergency manager for the Port of Seattle, calls "a disaster waiting to happen."[12] Not only does the artificial island straddle the Seattle Fault, but it also sits at the leading edge of the former tideflats, where the Duwamish River delta drops off into the deep water of Elliott Bay. During an earthquake, there is a "significant threat" of rupture along the delta front, which would cause it to slide into the bay, carrying terminals, docks, and cranes along with it.[13] If the front end of the island doesn't drop into deep water, the island will still sustain intense damages when it starts to shake, as it did in the Nisqually quake. Harbor Island also faces the triple threat of sea-level rise, storm surges, and flooding by the Duwamish River, each of which could swamp the island.

In response, the Port of Seattle has instituted long-range plans that take into account Harbor Island's challenges, says Joseph Gellings, a senior planner for the port. They are using up-to-date science and building codes to make plans for seismic issues as well as sea-level rise and short-term flooding events. "We believe that accommodation and protection strategies will suffice for at least the next fifty years," says Gellings. The port "must maintain its status as an engine of economic activity in King County," and port management has no plans to relocate, he adds.[14]

◆ ◆ ◆

As I stand on the observation deck of the Space Needle and look out over Seattle and Bertha's exit pit, the construction along the seawall, and the many new building projects below, I realize that Gellings's comments epitomize the modern reality of our landscape. We are a major city in a world where global trade has become routine. A successful and thriving Seattle relies on keeping open its trade routes, many of which converge on the made lands of the former tideflats. We cannot simply get rid of Harbor Island, move the port's facilities, or somehow fix the unconsolidated soils of the tidelands any more than we can rebuild Denny Hill. For better or worse, these areas have become integral to the city.

Just as early Seattleites had to adapt to what they found, we now have to adapt to what our predecessors left us. It will not be easy. We live in a far more complicated world with more rules and regulations, more people whose voices need to be heard, and vastly higher costs in completing large-scale engineering projects. Seattle may be on the edge of the continent, but it is no longer an isolated little city. We have to consider outside forces, both natural ones and those created by humans. These concerns do not mean that we won't be able to adapt. They do mean that, in whatever decisions we make, we will need to be more creative and more inclusive than we were in the past.

Seattle is a city of hills and valleys, ridges and bluffs, creeks and lakes, faults and landslides. It is a city governed by its geology and its topography. We will continue to worry about our surroundings and try to navigate around the city's myriad topographic challenges, but with one fundamental difference: future landscape change will be not about what we want to do but about what we have to do. In the end, though, the result will be the same. We will continue to shape the topography and it will continue to shape us.

Appendix

Volume of Dirt Moved in Seattle via Topographic Reshaping

SOURCE OF DIRT	VOLUME (yd^3)	SOURCE OF INFORMATION
Regrades (1911–28)	13,537,575	Bagley and Morse, minus their numbers indicating material removed during Denny Hill's fourth regrade (1908–11) and the Dearborn regrade
Regrades (1903–10)	1,085,643	Morse
Denny Hill regrades	11,113,025	Morse, Phelps, Moore, and "Regrading-North District"
Jackson Street regrade	3,347,883	Overstreet
Dearborn Street regrade	2,905,000	Overstreet
Locks region	1,661,400	Connole
Ship canal	4,000,000	*Seattle Times,* July 12, 1916
Duwamish waterway	7,400,000	*Seattle Times,* April 19, 1914
Harbor Island	7,000,000	*Seattle Times,* April 24, 1907*
Duwamish tideflats filling	21,736,069	Savidge
Great Northern Railroad line north through Interbay	1,250,000	*Seattle Times,* June 25, 1912
Total	75,036,595	

Sources: Clarence Bagley, *History of King County* (Seattle: S. J. Clark Publishing Company, 1929), 681–683; Susan Connole, Friends of the Ballard Locks, email correspondence with author, February 3, 2014; George Holmes Moore, "Heavy Regrading by Means of Hydraulic Sluicing at Seattle, Washington," *Engineering Record* 63, no. 13 (March 31, 1910): 355–358; Roy M. Morse, "Regrading Years in Seattle," in *Engineering Geology in Washington,* ed. R. W. Galter, Washington Division of Geology and Earth Resources Bulletin no. 78 (Olympia: Washington State Department of Natural Resources, 1989): 691–701; R. M. Overstreet, "Hydraulic Excavation Methods in Seattle: Detailed Construction Records of Some Features of the Jackson Street Regrade," *Engineering Record* 65, no. 18 (May 4, 1912): 480–483; Myra L. Phelps, *Public Works in Seattle: A Narrative History [of] the Engineering Department, 1875–1975* (Seattle: Seattle Engineering Department, 1978), 20; "Regrading-North District," [c. 1910], Local Improvement District 4818, Letters, Folder 3, Seattle Municipal Archives; Clark V. Savidge, *Thirteenth Biennial Report of the Commissioner of Public Lands to the Legislature of the State of Washington, October 1, 1912, to September 30, 1914* (Olympia: Frank M. Lamborn, 1915); *Seattle Times*, "Adding Twelve More Miles to Seattle's Water Front," April 19, 1914; *Seattle Times*, "Great Northern Lets Contracts for Work to Cost $1,250,000," June 25, 1912; *Seattle Times*, "Harbor Island Is to Be Filled at Once," April 24, 1907; *Seattle Times*, "Locks Closed and Lake Canal Is Filling," July 12, 1916.

*This is the only record I could find from a paper published during the period when Harbor Island was built. The *Seattle Times* published several later articles that provide a much larger number: 25 million cubic yards is the amount cited in E. B. Fussell, "Puget Sound Bridge Co. Has Changed 'Face' of Seattle," *Seattle Times*, September 27, 1950; and 24 million cubic yards, in Don Duncan, "Driftwood Diary," *Seattle Times*, May 7, 1967. I could find no information about this in the records of the Seattle Engineering Department or Port of Seattle.

Notes

INTRODUCTION

1. Clarence Bagley, *History of Seattle: From the Earliest Settlement to the Present Time* (Chicago: S. J. Clark Publishing, 1936), 1:20.
2. Nancy Seasholes, *Gaining Ground: A History of Landmaking in Boston* (Cambridge, MA: MIT Press, 2003), 2.
3. "Calls It a Germ," *Seattle Post-Intelligencer*, September 8, 1907. Hereafter *Seattle P-I*, even though the paper had various names during its early existence.
4. Welford Beaton, *The City That Made Itself: A Literary and Pictorial Record of the Building of Seattle* (Seattle: Terminal Publishing, 1914), 64.
5. David Buerge, "The Life and Death of the Black River," *Seattle Weekly*, October 16–22, 1985.
6. Coll Thrush, *Native Seattle: Histories from the Crossing-Over Place* (Seattle: University of Washington Press, 2007), 28–29.
7. The information on Boston comes from Seasholes, *Gaining Ground*.
8. Ibid., 423.
9. Craig E. Colten, "Chicago's Waste Lands: Refuse Disposal and Urban Growth, 1840–1990," *Journal of Historical Geography* 20, no. 2 (1994): 128.
10. Robin Amer, "How Has Chicago's Coastline Changed over the Decades?" WBEZ91.5, December 11, 2012, www.wbez.org/series/curious-city/how-has-chicago's-coastline-changed-over-decades-104328.
11. Colten, "Chicago's Waste Lands," 124.
12. Information on the railroad and movement of fill is from Matthew Morse Booker, *Down by the Bay: San Francisco's History between the Tides* (Berkeley: University of California Press, 2013). Sam Safran, of the San Francisco Estuary Institute, calculated for me the number of acres of fill. It is based on fill placed in open water (defined as shallow bay, deep bay, and channel) and tidal flats. Another thousand acres of tidal marshes, primarily along Mission Creek and Islais Creek, were also filled in. The total refers only to San Francisco County and so does not include the airport.

13 Michael Holleran, "Filling the Providence Cove: Image in the Evolution of Urban Form," *Rhode Island History* 48, no. 3 (August 1990): 81.

14 Robert Louis Stevenson and Lloyd Osbourne, *The Wrecker* (New York: Charles Scribner's Sons, 1905), 160–161.

15 Michael Chrzastowski, *Historical Changes to Lake Washington and Route of the Lake Washington Ship Canal, King County, Washington*, USGS Water Resources Investigation Open-File Report 81–1182 (Washington, DC: U.S. Geological Survey, 1981).

16 "Washing Seattle's Big Hills into the Sea," *Seattle P-I*, April 8, 1909. My estimated number is higher than the figure of 50 million cubic yards used by most historians, which I am pretty sure originates with Clarence Bagley's *History of King County*. His book came out in 1929 and does not include the final Denny Hill regrading. Nor does it include the Lake Washington Ship Canal, the straightening of the Duwamish Waterway, or the filling in of the Duwamish, all of which I include. Bagley writes, "As this figure [37,085,730 cubic yards] does not include private property and utility contract operations, the total yardage to date may be close to fifty million cubic yards." I have not added to my number Bagley's 13 million, because I have no way to verify that number, so my total is probably too low; but we will never know.

CHAPTER 1. GEOLOGY

1 Known as the Olympia beds, the tan-to-brown rocks were deposited more than twenty-five thousand years ago, before the Puget lobe reached Seattle, when the area was not experiencing an ice age and had a climate somewhat similar to our modern one, but with Columbia mammoths and giant sloths.

2 Conversation with Derek Booth, February 11, 2014.

3 Richard Galster and William T. Laprade, "Geology of Seattle, Washington, United States of America," *Bulletin of the Association of Engineering Geologists* 28, no. 3 (August 1991): 259.

4 This idea had been proposed in the 1960s by geologist Rocky Crandell, says Booth. "His work was overlooked. When we made our proposal, the reaction, if people stopped to think about this, was: 'Oh, yeah.'"

5 Booth has estimated that the average rate of water discharge out of the glacier was about a hundred thousand cubic feet per second, roughly equivalent to the volume of six Skagit Rivers. He further determined that the rivers flowing under the ice removed 240 cubic miles of earth. "Glaciofluvial Infilling and Scour of the Puget Lowland, Washington, during Ice-Sheet Glaciation," *Geology* 22 (August 1994): 695–698.

6 Renton Hill, also known as Second, is located at Seventeenth Avenue East and Madison Street. Profanity, or Yesler, is at the present location of Harborview Hospital. The name Profanity Hill comes from the words uttered by lawyers who had to ascend the hill to reach the old county courthouse that used to sit atop the hill.

7 These two hills are not considered part of the original seven because they were

not part of Seattle. Magnolia wasn't annexed to Seattle until 1891, and West Seattle's annexation occurred in 1907.

8 Conversation with Bill Laprade, March 5, 2014.

9 These numbers come from the University of Florida's former Panama Canal Museum, now only online.

10 Geologists use the term *group* to refer to a suite of related rocks. The Puget Group consists of the Renton Formation, Tukwila Formation, Tiger Mountain Formation, and Raging River Formation. The south Seattle hills are made of the Tukwila Formation.

11 Conversation with Brian Sherrod, February 14, 2014.

12 Conversation with Ralph Haugerud, February 20, 2014.

13 R. H. Thomson letter to John K. Brown, February 27, 1897, Record Group 2600–02, Box 1, Vol. 4, Page 428, Seattle Municipal Archives.

14 Technically, a hill such as this is slumping and not sliding, but such movement is still not a good sign. You can see evidence in the trees, too. On a slumping slope, the trees either tilt backward or curve up at the bottom.

15 Shannon and Wilson, Inc., *Seattle Landslide Study* (Seattle: Seattle Public Utilities, January 27, 2000), 1:26. The updated database for the city of Seattle has pushed the number of slides to more than fifteen hundred. "Seattle Landslide Database," in ibid., www.seattle.gov/dpd/landslidestudy/landslide%20study%20database.pdf.

CHAPTER 2. SEATTLE'S HISTORIC DOWNTOWN SHORELINE

1 There appears to be no original of this map still in existence. We do not know exactly when Phelps drew it, but he did include a building—the southern blockhouse—that was not built until March 1856. He apparently re-inked the map after 1879, when he became a commodore, the rank he held when he signed the drawing. Nor is it known when the map first appeared in print. The earliest rendition appears to be in Arthur Denny, *Pioneer Days on Puget Sound* (Seattle: Alice Harriman Company, 1908), which incorporates the map into a 1908 map of the city showing street layout and early land claims. As a stand-alone map, it next appears in Beriah Brown, "Seattle's First Taste of Warfare," *Town Crier*, December 15, 1917. The first time it appears with Phelps's writing is in Dorothy Fay Gould, *The Indian Attack on Seattle, January 26, 1856* (Seattle: Farwest Lithography and Print Company, 1932), 32.

2 Thomas Phelps, "Reminiscences of Seattle, Washington Territory, and the U.S. Sloop-of-War 'Decatur,' during the Indian War of 1855–56," *United Service: A Monthly Review of Military and Naval Affairs* 5 (1881): 696.

3 Ibid., 687, 692.

4 The name *Decatur* lives on in Seattle at Decatur School in north Seattle. The school was actually named for Stephen Decatur, but then so was the ship. You can also find a plaque at Third and Jefferson that commemorates the battle.

5 U.S. Coast Survey, Preliminary Survey of Duwamish Bay, W. T., United States

Coast Survey, 1854, Historic Map Cases, G4282.E42 P5 1854, University of Washington Library.

6 Built in 1935, the original bridge, or overpass, ran from Lenora Street over Alaskan Way to the second level of a warehouse at Pier 64–65. The overpass survived for six decades, until the Port of Seattle decided to redevelop the waterfront and removed the remaining section of the bridge over Alaskan Way. In order to provide access up to the Pike Place Market, the port left this short section of elevated concrete in place and added a stairway and elevator to reach the span from waterfront level.

7 Denny, *Pioneer Days*, 39.

8 My description of the bluff is based on discussions I had on June 28, 2012, and May 16, 2014, with Hugh Shipman, a coastal geologist with the Washington State Department of Ecology.

9 Dennis Lewarch et al., *Archaeological Evaluation and Construction Excavation Monitoring at the World Trade Center, Baba'kwOb Site (45KI456), Seattle, King County, Washington,* LAAS Technical Report no. 2002–15 (Gig Harbor, WA: Anthropological Archaeological Services, 2002.

10 "Ordinances of the Town of Seattle," *Seattle Weekly Gazette*, March 4, 1865. The ordinance passed on February 7, 1865. This is the first Ordinance No. 5. The second was passed in December 1869. The town of Seattle was incorporated in January 1865 but then unincorporated two years later. After reincorporation in December 1869 as the city of Seattle, it began a new numbering system.

11 John W. Reps, *Panoramas of Promise: Pacific Northwest Cities and Towns on Nineteenth-Century Lithographs* (Pullman: Washington State University Press, 1984), 1.

12 Lucile McDonald, "She Threw a Rock at the Daughter of Chief Sealth," *Seattle Times*, June 16, 1955. (Like the *P-I*, the *Seattle Times* had several names during its early existence; nonetheless I use the name *Seattle Times* throughout.)

13 The word *beachcomber* originated in the South Pacific in the 1840s as a derogatory term. Both Herman Melville and Robert Louis Stevenson used it. Speaking of beachcombers, Melville wrote in *Omoo*, "Their reputation is a bad one."

14 "The Beach-Combers," *Seattle P-I*, March 15, 1891.

15 The first quote is from Rose Simmons, "Old Angeline, the Princess of Seattle," *Overland Monthly* 20 (November 1892): 506. The second is from "Trouble in Shantytown," *Seattle P-I*, December 23, 1900.

16 "The Front Street Wharf," *Seattle P-I*, January 16, 1877.

17 The information about the Woodward Grain House comes from Paul Dorpat's book *Seattle Waterfront: An Illustrated History* (Seattle: Seattle City Council, 2006), 35.

18 "City Council," *Seattle P-I*, July 8, 1876; "City Council," *Seattle P-I*, July 10, 1876.

19 "A Good Work," *Seattle P-I*, August 10, 1876.

20 "Street Grading," *Seattle P-I*, September 19, 1876.

21 "Improvement," *Pacific Tribune Weekly*, October 7, 1876.

22 "Grading," *Seattle P-I*, November 22, 1876.

23 "Front Street," *Seattle P-I*, March 28, 1877.

24 "The Seattle Street Grades," *Daily Pacific Tribune*, September 27, 1876.

25 "Street Grading," *Seattle P-I*, October 15, 1876.

26 Information on the core samples is drawn from a letter to M. M. Lwin, Washington Department of Transportation, from Tony Allen, state geotechnical engineer, June 27, 1996.

27 Conversation with Bob Kimmerling, July 2012.

28 One ingredient missing from the cores is gold, from the Klondike gold rush. Surely some bags, nuggets, or flakes must have slipped overboard from the ships that dropped off miners who were headed to Seattle's assay office. But none has been sighted, probably because the gold is too fine.

29 The Polson Building next door doesn't suffer from the tides. Its more extensive basement means that its pilings are deeper underground and, as a result, never, or very rarely, exposed during tidal fluctuations.

30 Coll Thrush, *Native Seattle: Histories from the Crossing-Over Place* (Seattle: University of Washington Press, 2007), 229.

31 Ibid., 102.

32 The term *Skid Road* came from the logs skidded down the steep upper parts of Mill Street to Yesler's mill on the waterfront. The area south of Yesler became known as "Skid Road," a name that later morphed into "Skid Row" and was transferred to a less-than-desirable part of town.

33 This is the only map known that includes this specific honor for Thomas Piner, a quartermaster for Wilkes. Wilkes also named a bay in Antarctica for Piner, as well as Point Piner on Vashon Island. Both of those names have persisted to the present.

34 Historian Greg Lange believes that the man who invented Seattle Underground Tours, Bill Speidel, coined the name Denny's Island. Lange could find no earlier mention of the term than Speidel's book *Sons of the Profits*, written in 1967 (Greg Lange, email correspondence with the author, November 21, 2012). Paul Dorpat agrees with Lange, writing that if the term doesn't appear earlier than Speidel's book, assertions about the historical veracity of the name Denny's Island "should be approached with caution and treated with doubt." Paul Dorpat, email correspondence with the author, November 20, 2012.

35 Clarence Bagley, *History of Seattle: From the Earliest Settlement to the Present Time* (Chicago: S. J. Clark Publishing, 1936), 1:21.

36 Before 1875, there were two "official" reference points in the city. One was an *X* cut in a boulder at the intersection of the center lines of Mill Street and Commercial Avenue (now First Avenue South). The other was a nail-centered hub (a surveyor's marker) driven a little below the surface at Commercial Avenue and Main Street.

37 Six years later the city revised this number to 9.6 feet below the top of the granite step in the bank's doorway. I haven't been able to determine why this was done. In looking at old photos of the bank, I wonder if they simply removed the lower step and thus had to change the datum point.

38 "The Council's Action," *Seattle Times*, June 13, 1889.

39 In 1889, the city assessor listed the population as 33,500. According to *The Oregonian's Handbook of the Pacific Northwest* (Portland: Lewis and Dryden Printing Company, 1894), the Territorial Census of Washington listed the population as 26,740.

40 David Buerge, *Seattle in the 1880s* (Seattle: Historical Society of Seattle and Puget Sound, 1986), 55.

41 "Two Duties," *Seattle P-I*, June 12, 1889.

42 Although the city controlled the project, building owners had to pay for and erect the retaining walls in front of their properties. This led to a variation in materials used. Walls were made of concrete, brick, sandstone, or rubble (stacked, broken concrete blocks held together by mortar), or combinations of these.

43 Initially developed in the 1840s as deck lights for ships, vault lights were popular in cities, including New York, Chicago, and Boston, from the late 1800s through the 1930s, when electric light became more popular.

44 Conversation with Brandy Rinck, July 13, 2013.

45 Conversation with Hugh Shipman, June 28, 2012.

CHAPTER 3. FILLING IN THE DUWAMISH RIVER TIDEFLATS

1 "Railroad Work Begun," *Puget Sound Dispatch*, May 7, 1874.

2 There is no record of exactly where the Seattleites gathered; but based on my March 1, 2013, discussion with the Seattle train history expert Kurt Armbruster, as well as on newspapers of the day and later books by early Seattle residents, this seems as likely a spot as any other.

3 Telegram to A. A. Denny, Seattle, July 14, 1873, Asahel Curtis Photo Company Photographs, University Libraries, University of Washington, Digital Collections, http://digitalcollections.lib.washington.edu/cdm/ref/collection/curtis/id/1069.

4 "A Railroad Basis," *Puget Sound Dispatch*, July 31, 1873.

5 "Seattle's Opportunity," *Puget Sound Dispatch*, July 24, 1873.

6 "Ordinance No. 44," *Puget Sound Dispatch*, August 21, 1873.

7 Posters around Seattle included admonitions such as "Come all ye adults of mankind/Nor let there be none left behind." One wag added in blue pencil: "Come each of you and bring your tools / And work away like ___ ___ fools / Don't beat the devil about the bush / But skoot the road to Mox-la-push." "By the Way," *Seattle Times*, July 21, 1893.

8 "Railroad Work Begun," *Puget Sound Dispatch*, May 7, 1874.

9 Ibid.

10 The best place to get a true feeling for the tideflats is the Nisqually National Wildlife Refuge. You can walk on an elevated boardwalk out over the hundreds of acres of tideflats, which, as in Seattle, are fed by a river and surrounded by forested slopes.

11 Conversation with Warren King George, May 13, 2014.

12 Noted in an 1868 survey by San Francisco engineer Theodore A. Blake, as cited in

Watson Andrews Goodyear, *Coals Mines of the Western Coast of the United States* (San Francisco: A. L. Bancroft, 1877), 112.

13 The barge trips had their own trials. In January 1875, a barge towed by the stern-wheeled steamer *Chehalis* sank in a storm on Lake Washington. The eighteen coal cars formerly on the barge now rest in two hundred feet of water south of the State Route 520 bridge. Divers report that most of the cars are upright on the bottom, still holding coal.

14 Quote from "A Gala Day," *Puget Sound Dispatch*, March 28, 1872. The Duwamish took sixteen river miles to travel the eight straight miles from its confluence with the Black to tidewater. Snags, shoals, eddies, and low-hanging trees made the route slow going and dangerous for barge travel. And then the empty barges had to get back upriver to the mines.

15 Pile drivers used on Puget Sound in the 1890s struck seventy blows per minute.

16 E. I. Carpenter, "Improvement of the Swinomich Slough, Washington," *Annual Report of the Chief Engineers of the United States Army to the Secretary of War for the Year 1895* (Washington, DC: Government Printing Office, 1895), pt. 5, p. 3438.

17 James Elverson, "Ocean Shells," *Seattle Times*, October 11, 1908.

18 "The Teredo," *Scientific American*, 64, no. 17 (April 25, 1891): 266.

19 I have based my argument on Alexander MacDonald, "Seattle's Economic Development, 1880–1910" (PhD diss., University of Washington, 1959).

20 "The Opening—a Success," *Seattle P-I*, March 8, 1877.

21 *Report of the Secretary of the Interior; Being Part of the Message and Documents Communicated to the Two Houses of Congress at the Beginning of the Second Session of the Forty-Eighth Congress* (Washington, DC: Government Printing Office, 1884), 2:600.

22 MacDonald, "Seattle's Economic Development, 1880–1910," 77. The figure of one thousand homes comes from the city directory. *The City Directory 1884–85 of Seattle, Washington* (Seattle: Industrial World, 1884), 19.

23 "Local," *Seattle P-I*, July 26, 1882.

24 The original source for this information appears to be J. Willis Sayre, *This City of Ours* (Seattle: Seattle School District, 1936), 69. Sayre, who was a theater critic and journalist, wrote several books about Seattle history. Each is quirky and provides firsthand stories about early Seattle. His list of cities shows up repeatedly in books about Seattle. There is no reason to doubt Sayre's observation of where the ballast originated, but there are no references to support it, either. In 2014, during construction of a shaft for repairing the tunnel-boring machine Bertha, archaeologists excavated exploratory pits and found pieces of Ballast Island, including mixed sand and slag, as well as pieces of schist, granite, and sandstone. The largest were boulders, and many were the size of a small soccer ball. One of the archaeologists happened to be in San Francisco soon after the dig and visited Telegraph Hill. He told me that the quarry rock extracted there looked exactly like what they had excavated in Seattle.

25 Coll Thrush, *Native Seattle: Histories from the Crossing-Over Place* (Seattle: University of Washington Press, 2007), 85.

26 An apocryphal story from the era refers to one prostitute, Sweet Alice Duval, who capitalized on the location by having a trapdoor in her room, where she could kill and rob her clients and drop them into the water below. James Stevens, "The Natural History of Seattle," *American Mercury* 27, no. 108 (December 1932): 402–409.

27 Pier 46 again became a home for the homeless in the early years of the twenty-first century. They were eventually kicked out when the Port of Seattle found tenants to pay for the space.

28 The most widely used of these, Valentine scrip, originated from a claim by Thomas B. Valentine on property in California, subsequently taken from him by the U.S. government. In return, Valentine was awarded Special Certificates of Location, or scrip, which gave him 13,316 acres of, according to a statement on the scrip itself, "unoccupied and unappropriated public lands." He was entitled to 332 certificates, each of which could be used to acquire forty acres of land. What made the scrip certificates particularly valuable was that they could be transferred or sold to anyone. Early investors would buy scrip certificates for as little as $6 per acre, or $240 total. They could then redeem the certificate for land, which they hoped to sell for more than $6 per acre, or they could resell the certificate. By the time the speculation ended in a law case, the price per acre had risen to $200 and sometimes significantly higher, which meant that a savvy speculator could have turned a $240 investment into $8,000. Valentine scrip was used in seventeen states, from California to Iowa. This information comes from several articles: "They Want the Tide Flats," *Seattle P-I*, January 10, 1889; "Jumping Tide Lands," *Seattle P-I*, September 26, 1889; "Scrip Men Appeal," *Seattle P-I*, November 1, 1889. For additional background on Valentine scrip, see Robert Lee, "Valentine Scrip: The Saga of Land Locations in South Dakota Territory Originating from a Mexican Land Grant," *South Dakota History* (South Dakota State Historical Society) 2, no. 3 (1972).

29 Testimony of lawyer N. Soderberg before the Tide Lands Committee, *Senate Journal of the First Legislature of the State of Washington* (Olympia, WA: O. C. White, 1890).

30 "Tideflats Are in Demand," *Seattle P-I*, July 5, 1889. In a curious bit of chutzpah, the carpenters building the shacks went on strike demanding higher wages for their illicit work.

31 Burke to Carrie Allen, May 1, 1888, Box 20, Accession 1483-002, University of Washington Special Collections.

32 Denny had first filed a claim in 1869, looking for plumbago (reddish lead ore), which he had heard was used by the Snoqualmie tribe to make face paint. In 1882, he and several others filed more extensive claims on what is now called Denny Creek. Denny would eventually lease these ore claims to Peter Kirk, the British owner of Moss Bay Hematite Iron and Steel Works in England, who planned to build a massive steel plant on the east side of Lake Washington, at the present site of Kirkland. See chapter 4 for additional information on Kirk.

33 "To Connect with the Canadian Railroad," *Seattle P-I*, April 15, 1887.

34 Ordinance No. 804 passed on January 25, 1887. Ordinance No. 806 passed on January 27.

35 "Railroad Talk," *Seattle Times*, March 6, 1887.

36 Sanborn Map and Publishing Company, *Seattle, Wash. Ter. 1888* (New York: Sanborn Map and Publishing Company, 1888).

37 Based on 1884 and 1888 Sanborn Fire Insurance Maps. Sanborn Map and Publishing Company, *Seattle, W. T. July 1884* (New York: Sanborn Map and Publishing Company, 1884); Sanborn Map and Publishing Company, *Seattle, Wash. Ter. 1888*.

38 Letter to Crawford, February 13, 1887, cited in Robert C. Nesbitt, *He Built Seattle: A Biography of Judge Thomas Burke* (Seattle: University of Washington Press, 1961), 117.

39 This analysis is based on MacDonald, "Seattle's Economic Development, 1880–1910."

40 "The Fire," *Seattle P-I*, June 8, 1889.

41 *Seattle Times*, August 4, 1889.

42 "The Water Front," *Seattle Times*, June 24, 1889.

43 By 1910, the Great Northern; Union Pacific; Northern Pacific; and Chicago, Milwaukee, Saint Paul, and Pacific Railroads all had transcontinental lines reaching Seattle.

44 In 1903, the newspaper the *Commonwealth* led a campaign to clean up Railroad Avenue, claiming that lighting and policing were practically nonexistent, that trains didn't follow city ordinances, and that the sanitary conditions under the trestle were unbearable, disease-breeding, and frightful.

45 "Crichton Asks for a Sea Wall," *Seattle Times*, April 7, 1908.

46 In 1909, city engineer Thomson proposed not merely a seawall but an extension of the waterfront by as much as fifteen hundred feet into Elliott Bay. Thomson's grand scheme would have created more than three hundred acres of made land and cost $17.5 million. Although the mayor initially endorsed it, the plan did not go further than paper. Thomson makes no mention of the plan in his autobiography. The plan was described in "Thomson Submits Plan for Seawall," *Seattle Times*, December 25, 1909.

47 Alan Hynding, *The Public Life of Eugene Semple: Promoter and Politician of the Pacific Northwest* (Seattle: University of Washington Press, 1973), xi.

48 The U.S. government became trustee of the tidelands in 1846 when it acquired the Oregon Territory through the Oregon Treaty, which clarified the U.S.–Great Britain boundary at the forty-ninth parallel. Moreover, in the Treaty of Point Elliott, signed on January 22, 1855, the federal government gained the rights to many Native lands around Puget Sound. The tribes, however, retained the right to hunt, gather roots and berries, and fish at their "usual and accustomed grounds and stations," which raises the question of whether any alteration of tidelands was legal.

49 For a thorough discussion of the debate over the tideflats and tidelands at the time of statehood, see Charles Wiggins, *The Battle for the Tidelands in the Constitutional Convention*, pts. 1, 2, and 3, which ran in the *Washington State Bar News* in March, April, and May 1990.

50 Eugene Semple, "Preliminary Prospectus (Private and Confidential) of Proposed Improvement in Seattle Harbor, State of Washington," undated, Eugene Semple

Papers, Box 10, Folder 2, Page 6, Accession 0532–001, University of Washington Special Collections.

51 Semple's plan was based closely on a harbor scheme developed by civil engineer Virgil Bogue. In 1897, assistant city engineer George Cotterill modified Bogue's plat plan for the tidelands, particularly in regard to piers. Before Cotterill's replat, most piers were simply extension of the roads, which meant they ran perpendicular to most of the train tracks. This created havoc where the shoreline curved and piers converged. With Cotterill's plan, all piers would be parallel and run east-west. Although pier owners were slow to accept the plan, they eventually followed it, leading to the pier alignment we have today. If you want to test a friend's powers of observation, ask him or her which way the piers run. Most people will say northwest-southeast—and the piers do seem to run that direction, because the streets of downtown are among the few in Seattle that do not run north-south.

52 "An Auspicious Event," *Seattle Times*, July 29, 1895.

53 Thankfully, wrote one reporter, that nasty "old Sol" was hidden behind the clouds and no one would get too hot.

54 "Digging the Ditch," *Seattle P-I*, July 30, 1895. The paper stated that the fish was a "bullhead." One fish commonly called a bullhead at that time, and still identified as such today, is the sculpin.

55 "Unsafe Pile-Driving," *Oregonian*, May 21, 1890, described two, two-story brick buildings on piles that collapsed during construction. Apparently the piles had not been driven deep enough and had begun to settle under the weight of the structures.

56 "Make Plans to Beautify," *Seattle Times*, March 26, 1905.

57 The complex of buildings eventually morphed into the home of Rainier Beer. I have also been told that, as late as the 1980s, there was a job at the Burlington Northern tracks near the brewery that was known as the "shoreline job," in reference to the area's former location along the shoreline.

58 "Large Areas of Seattle Tide Lands to Be Reclaimed," *Seattle Times*, October 12, 1901.

59 I have not been able to determine if the Canal Waterway was actually completed. It appears to have been started, and it is shown on the 1912 Baist map and the 1904–5 Sanborn map. It is also mentioned in the newspaper, or at least its filling in is mentioned, but there are no photographs of it. Paul Dorpat raises a good point, asking why the canal would be dug before completion of the cut through Beacon Hill. Why then are there articles in the *Seattle Times* describing the filling in of a canal at this location? It's one of Seattle's little mysteries. Paul Dorpat, email correspondence with author, January 30, 2014.

60 Conversation with Bob Norris, July 2013.

CHAPTER 4. REPLUMBING THE LAKES

1 Clarence Bagley, *History of Seattle: From the Earliest Settlement to the Present Time* (Chicago: S. J. Clark Publishing, 1936), 1:371.

2 Matthew Klingle, *Emerald City: An Environmental History of Seattle* (New Haven, CT: Yale University Press, 2007), 62.

3 Gordon Dodds, *Hiram Martin Chittenden: His Public Career* (Lexington: University of Kentucky Press, 1973), 130.

4 "Seattle's Ship Way, Dream of Years, Opened," *Seattle Times*, July 4, 1917.

5 "Lake Washington, Washington," published by the U.S. Coast and Geodetic Survey, 1905, Early Washington Map Viewer, WSU Libraries Digital Collections, Washington State University, http://kaga.wsulibs.wsu.edu/zoom/zoom.php?map=uw120.

6 The earliest population data for Renton—488 people—is from 1880 and includes census numbers for Duwamish and Talbot, areas south and west of Renton proper. This information comes from Elizabeth Stewart, director of the Renton History Museum.

7 Howard McKinley Corning, *The New Washington: A Guide to the Evergreen State* (Portland, OR: Binfords and Mort, 1941), 322–323.

8 "Renton Will Celebrate," *Renton Herald*, May 30, 1912.

9 Historically, the Green River flowed into the White River in Auburn, and the two continued as the White to its confluence with the Black. Floods in 1906, however, changed the course of the White, which then drained, and still drains, into the Puyallup River. The Green kept its course and became the outflow for the Black, until the disappearance of the Black in 1916—which is why the Green changes name for no apparent reason and becomes the Duwamish.

10 When the Cedar River flooded in 1911, it was said that you could paddle a boat from Kent to Renton, a distance of seven miles or so.

11 David Buerge, "The Life and Death of the Black River," *Seattle Weekly*, October 16–22, 1985.

12 My analysis is based on a *Kroll Atlas of Seattle and Supplements* (Seattle: Kroll Map Company, undated) owned by the Renton History Museum. It contains several maps of Renton, which include outlines of the "Abandonded Black River Channel." The 1900 USGS topographic map of Tacoma shows a slightly different route for the Black, one in which it flows under the train tracks a bit farther to the west.

13 Hiram M. Chittenden, *Report of an Investigation by a Board of Engineers of the Means of Controlling Floods in the Duwamish-Puyallup Valleys and Their Tributaries* (Seattle: Hanford and Lowman, 1907), 16.

14 Buerge, "Life and Death," 49. Conversation with Warren King George, May 13, 2014.

15 It is not known when these species reappeared in the lake or whether they are native populations. Roughly 10 to 15 percent of salmon that return to a specific location are strays that originally hatched some other place. At present, adult salmon use both the ladder and the locks to enter the lake.

16 This section is based on conversations with Sally Abella, who did critical research in Lake Washington with Tommy Edmonson, the scientist most responsible for telling the story of the lake.

17 Conversation with Charles "Si" Simenstad, April 1, 2013.

18 Smith quoted by Lucile McDonald, "Early Developments in Southeast Seattle," *Seattle Times*, January 15, 1956. In 1912, when the Olmsted Brothers landscape architecture firm was working on its plans for Seattle's park system, some locals agitated for a canal through the isthmus so that tour boats could pass through. They stated that the steamers would be a boon to "poor people [who] had no means of getting to Bailey Peninsula . . . without walking a great way," noted Olmsted employee James Dawson in a report to the parks department on April 5, 1912.

19 Conversation with Kurt Fresh, September 11, 2013.

20 Mail would be tossed twice a day from the SLS&E. The postmaster's dog was trained to go and fetch the mailbag.

21 Statistics on output from "Yesler Mill Burned," *Seattle Times*, September 17, 1895; Lucile McDonald, "How Lake Washington Changed," *Seattle Times*, October 2, 1955.

22 This description comes from George B. Rigg, *Peat Resources of Washington*, Bulletin 44 (Olympia, WA: Division of Mines and Geology, State of Washington, 1958), 259. It does not refer specifically to this site but to one nearby.

23 Information on Japanese and Italian truck farmers driving through neighborhoods comes from an interview with Dan Evans by Judith Thornton, *Waterway 1: Preserving a Village Green* (Olympia, WA: Washington Heritage Register, Washington State Department of Archaeology and Historic Preservation, August 2010).

24 Greg Bishop, Wayne L. Turnberg, Karen Van Dusen, *Abandoned Landfill Study in the City of Seattle* (Seattle: Seattle–King County Department of Public Health, July 30, 1984), 5.

25 A 1915 Department of Health and Sanitation report contained the following sentence. The report pinpointed "an almost unlimited number of places within our city limits which can be beautified by means of sanitary fills."

26 Interbay, 55 acres; Genesee, 26 acres; near Harbor Island, 20 acres; Green Lake, 100 acres; Miller Street, 30 acres. Acreage based on numbers reported in Bishop, Turnberg, Van Dusen, *Abandoned Landfill Study in the City of Seattle*.

27 Archibald Powell, "The Proposed Lake Washington Canal, Seattle, Wash: A Great Engineering Project," *Engineering News* 63, no. 1 (January 6, 1910): 15.

28 Lucile McDonald, "Early Days of Kirkland, Houghton," *Seattle Times*, October 23, 1955.

29 Lucile McDonald, "Juanita Once Was Named Hubbard," *Seattle Times*, October 16, 1955.

30 Dorris Forbes Beecher and Delia Blakeney Durr, interviewed by Lilly Mae Anderson and Lucile McDonald, August, 17, 1978. Transcript owned by Eastside Heritage Center, Seattle.

31 Corning, *The New Washington*, 323, 324.

32 Letter to Lucile McDonald, undated, William S. Lagen Papers, Box 1, Folder "Historical features," Accession 2292–001, University of Washington Special Collections.

33 These include little-known names such as Sill, Lucerne, Peterson, Curtis, Northup, Burrows, and Dabney, as well as ones that are widely known, including Newcastle, Mercer, and Medina.

34 Conversation with Loita Hawkinson, August, 7, 2013.

35 "Dream Comes True as Steamer Osprey Races into Lake Washington," *Seattle Times*, July 4, 1918.

36 *Hewitt-Lea Lumber Company v King County*, 257 U.S. 622, 42 S. Ct. 186, 66 L.Ed. 402.

37 Lucile McDonald, "Heavy Industries Were Planned for the Lake's Southeastern Shore," *Seattle Times*, November 13, 1955. For example, McDonald noted that Factoria was named for "the new smokestack city" expected to pop up early in the first decade of the twentieth century.

38 Bagley, *History of Seattle*, 371.

39 R. H. Thomson to John F. Miller, December 1, 1908, *Annual Report, City Engineer's Office*, 1908, Record Series 1802-G2, Box 1, Folder 1, Page 18, Seattle Municipal Archives.

40 R. H. Thomson, *That Man Thomson* (Seattle: University of Washington Press, 1950), 85.

CHAPTER 5. REGRADING DENNY HILL

1 R. H. Thomson, *That Man Thomson* (Seattle: University of Washington Press, 1950), 10.

2 William H. Wilson, *Shaper of Seattle: Reginald Heber Thomson's Pacific Northwest* (Pullman: Washington State University Press, 2009), 39.

3 Thomson, *That Man Thomson*, 13.

4 R. H. Thomson, *Annual Report, City Engineer's Office*, 1908, Record Series 1802-G2, Box 1, Folder 1, Pages 18–19, Seattle Municipal Archives.

5 R. H. Thomson to John F. Miller, December 23, 1909, *Annual Report, City Engineer's Office*, 1909, Record Series 1802-G2, Box 1, Folder 2, Seattle Municipal Archives.

6 Victor J. Farrar, "History of the University," *Washington Alumnus* (November 1920): 4–5, 13, 16.

7 Jeffrey Karl Ochsner and Dennis Alan Anderson, "Architecture for Seattle Schools, 1880–1900," *Pacific Northwest Quarterly* 83, no. 4 (October 1992): 131.

8 "Grading a Thoroughfare," *Seattle P-I*, August 7, 1908.

9 Ordinance No. 571, which passed on July 9, 1884, converted the land that David and Louisa Denny had donated to the city in 1864 for a cemetery into a public park. Originally known as Seattle Park, it was later renamed Denny Park.

10 Thomas Prosch, "A Chronological History of Seattle from 1850 to 1897" (unpublished manuscript, Seattle, 1901).

11 Sophie Frye Bass, *When Seattle Was a Village* (Seattle: Lowman and Hanford, 1947), 127.

12 George W. Baist, *Baist's Real Estate Atlas of Surveys of Seattle, Wash.* (Philadelphia: Baist, 1905). The number is based on my count of the blocks shown on the maps.

13 Aaron Raymond, "Denny Regrade (1893–2008): A Case Study in Historical GIS"

(master's thesis, University of Washington, 2009), 40. His thesis excludes the area west of Third Avenue.

14 Thomson, *That Man Thomson*, 14, 85.

15 R. H. Thomson, *Seattle Regrades*, undated, Box 13, Folder 15, Acquisition 0089-01, University of Washington Special Collections, 11.

16 Wilson, *Shaper of Seattle*, 212–213.

17 Thomson, *Seattle Regrades*, 4.

18 The position of city engineer was created by the Freeholders City Charter in 1890. The first to fill it was Albro Gardner, who lived on Denny Hill.

19 Front Street was renamed First Avenue on December 23, 1895, when Ordinance No. 4044 passed. This ordinance also changed many numbered streets to numbered avenues, in addition to changing the names of more than three hundred other streets.

20 "Grading a Thoroughfare," *Seattle P-I*, August 7, 1898.

21 C. T. Conover, "Just Cogitating: James A. Moore Had a Spectacular Career," *Seattle Times*, September 26, 1954.

22 Walter V. Woehlke, "Potlatch Town," *Sunset* 29, no. 1 (July 1912): 10.

23 Thomson's biographer notes that Thomson was a huge fan of this drawing.

24 Thomson's assistant had another idea for reducing costs. He proposed building escalators capable of carrying horses and their loads uphill. "Shall Seattle's Hills Be Equipped with Escalators?" *Seattle P-I*, July 23, 1905.

25 "Mystery Is Still Unsolved," *Seattle Times*, August 20, 1905.

26 At one point, Moore planned to tunnel under his hotel and build elevator access up into the hotel. He wasn't the first to propose a tunnel. In 1890, during the original construction, the developer of what was then the town of Fremont, L. H. Griffith, tried to get permission to tunnel under Denny Hill in order to build a rail line out to his new development.

27 Thomson, *That Man Thomson*, 90.

28 Matthew Klingle, *Emerald City: An Environmental History of Seattle* (New Haven, CT: Yale University Press, 2007), 114.

29 C. C. Closson, "Seattle's Regrade Projects: A Letter to the Editor of the Town Crier," *Town Crier*, August 26, 1911.

30 Hawley had a few troubles with the material he dumped in Elliott Bay, because the dirt would slide and block docks that then had to be dredged by the owners. Figures are from a document titled "Regrading-North District," [c. 1910], Local Improvement District 4818, Letters, Folder 3, Page 16, Seattle Municipal Archives.

31 The hill on the east side of Ninth Avenue was so steep that during a cold spell in January 1907, seven horses had to be killed after they slid on the ice. "Fifty Horses Are Killed on Icy Streets," *Seattle Times*, January 9, 1907.

32 Thomson, *Seattle Regrades*, 3.

33 "Two Tunnels Are Suggested," *Seattle Times*, April 10, 1904.

34 Thomson, *Seattle Regrades*, 4.

35 Details are from a document titled "The Jackson Street Regrade," [c. 1911], Local Improvement District 1213, Letters, Fiche 1, Seattle Municipal Archives.

36 "The Re-grading of Seattle, Washington II," *Engineering Record* 57, no. 20 (May 16, 1908): 638.

37 Total water usage for the project was 10,095,179,594 gallons, or enough water to fill 15,300 Olympic-size swimming pools.

38 "Jackson Street Regrade," 11.

39 Larger boulders, though, some up to ten feet in diameter, were not broken up but were taken away and used in terraces and walls.

40 This is based on a four-cubic-foot wheelbarrow model, which is the size typically used for home and garden.

41 R. H. Thomson credits William H. Lewis, of Lewis and Wiley, with introducing the technology to Seattle during their work on Jackson Street; but hydraulic hoses were in use as early as 1897. Thomson, *Seattle Regrades*.

42 "Westover Terraces among Country's Greatest Projects," *Sunday Oregonian*, October 19, 1913.

43 "Washing Seattle's Big Hills into the Sea," *Seattle P-I*, April 8, 1909.

44 The report on the project stated that it was "practically finished in December," but that the official paperwork wasn't accepted by the Board of Public Works until March 1910.

45 R. M. Overstreet, "Hydraulic Excavation Methods in Seattle," *Engineering Record* 65, no. 18 (May 4, 1912): 482.

46 Dearborn Street illustrates a problem with regrading in Seattle. In the first decade after completion of the bridge, two landslides of the exposed soil on the sides of the man-made canyon severely damaged the span.

47 Overstreet, "Hydraulic Excavation," 481.

48 Thomson, *Seattle Regrades*, 7.

49 "Denny Hill Tax Represents Equity," *Seattle Times*, May 12, 1907.

50 Conversation with William Wilson, March 19, 2014.

51 "Regrading-North District," 15.

52 This is a composite sketch based on the 1900 census, articles in the *Seattle Times*, and *Polk's City Directories* from 1901 to 1908.

53 *A Volume of Memoirs and Genealogy of Representative Citizens of Seattle and County of King, Washington* (New York: Lewis Publishing, 1903), 496.

54 Ward quoted by Margaret Pitcairn Strachan, "Early-Day Mansions," *Seattle Times*, July 15, 1945. Totally random, useless trivia: the son-in-law, William A. Dickey, is the man credited with naming Mount McKinley.

55 "November Realty Business Treble That of November, 1904," *Seattle Times*, December 3, 1905.

56 The original Cedar River pipeline provided 23 million gallons per day. A second pipeline, completed in 1909, provided an additional 40 million gallons per day.

57 I have not been able to determine who chose to waste the fill, or why. The *P-I* of April 8, 1909, notes that negotiations on selling the earth for fill "came to naught" but does not say who was involved in these discussions. The same issue would arise in the final regrading of Denny.

58 GS&C&S did not send all the Denny dirt into Elliott Bay. Toward the end of the

project, in October 1910, they also transported dirt inland to fill in the valley where Westlake Boulevard was being constructed. "Last Earth Being Dumped into the Bay," *Seattle Times*, October 24, 1910.

59 "Massive Buttes Monument to Anti-Regrades," *Seattle P-I*, January 10, 1910.

60 "Two of Tallest Spite Humps to Be Carted Away," *Seattle Times*, September 28, 1910.

61 "How to Move a House," *American Agriculturist* 32 (November 1873): 417; Owen Bernard Maginnis, *How to Move a House; How to Frame a House: Or House and Roof Framing*, 7th ed. (New York: William T. Comstock, 1914); David Stevenson, *Sketch of the Civil Engineering in America*, 2nd ed. (London: John Weale, 1859).

62 The Rinehart house ended up at 515 Thomas Street, about a half mile north. In 1930, the house traveled again, this time to allow the final regrading of Denny Hill. The owner, Belle Shepherd, moved the house off her site, had the mound it formerly rested on leveled, and then moved the house back. Shepherd rented out rooms in the house, which she called the Ruth Apartments. The apartments survived until 1952, when they became the property of the State Employment Security Department. In 1965, the building was leveled for a parking lot.

63 Municipal Plans Commission, *Plans of Seattle: Report of the Municipal Plans Commission Submitting Report of Virgil G. Bogue, Engineer* (Seattle: Lowman and Hanford, 1911), 131.

64 Ibid., 16, 17, 21.

65 Thomson, *Seattle Regrades*, 11.

66 "Thomson's Arrival and Retirement," *Seattle Times*, December 21, 1911.

67 Many of these handsome structures still exist, including the Pathé Building, built in 1922. The exchanges were distribution centers, where cinema owners could inspect films and decide if they wanted to show them in their theaters.

68 "Denny Hill at Last to Go," *Seattle Times*, February 17, 1927.

69 Schools in areas such as Laurelhurst, Green Lake, and Leschi had retention rates of around 80 percent.

70 Andrew W. Lind, *A Study of Mobility of Population in Seattle*, University of Washington Publications in the Social Sciences, vol. 3, no. 1 (Seattle: University of Washington Press, 1925), 16, 18, 23, 54–55.

71 One part of Denny School still exists: the piano donated to the school by David Denny. It still plays and is located on the second floor of Pioneer Hall in Madison Park.

72 "Steeple Tower on Denny Hill Falls to Earth," *Seattle Times*, March 30, 1929.

73 Another element of Denny Hill might still exist. When workmen cut down the trees on the hill, the wood was salvaged in October 1929 and made into handles for garden tools, which were sold to benefit Children's Orthopedic Hospital.

74 "Denny Hill Gone: Mayor Will Scoop Up Last of Earth," *Seattle Times*, December 9, 1930; "Mayor Finishes Denny Regrade," *Seattle Times*, December 10, 1930.

75 "Good-Bye Denny," *Seattle P-I*, December 11, 1930.

76 Curt Brownfield, Christian's great-grandson, told me, "Yeah, we owned the University District." Conversation, March 13, 2014.

77 Thomson, *That Man Thomson*, 92.

78 Clarence Bagley, *History of Seattle: From the Earliest Settlement to the Present Time* (Chicago: S. J. Clark Publishing, 1936), 1:360.

79 "Washington Ball Fitting Farewell," *Seattle P-I*, May 7, 1906.

80 Roger Sale, *Seattle: Past to Present* (Seattle: University of Washington Press, 1976), 76.

81 Janice L. Reiff, "Urbanization and the Social Structure: Seattle, Washington, 1852–1910" (PhD diss., University of Washington, 1981).

82 Sale, *Seattle*, 77.

83 V. V. Tarbill, "Mountain-Moving in Seattle," *Harvard Business Review* 8, no. 4 (July 1930): 482.

CHAPTER 6. WE SHAPE THE LAND AND THE LAND SHAPES US

1 Ralph Finke, *Alaskan Way Viaduct: Report on Preliminary Plans* (Seattle: City of Seattle Department of Engineering, June 1947).

2 The viaduct ultimately cost about $10 million, with the state covering 25 percent, the federal government 25 percent, and the city the remaining half. *Alaskan Way Viaduct and Battery Street Tunnel, HAER WA-184* (Washington, DC: Historic American Engineering Record, undated), 13, 14. In 2013 dollars, $10 million would be worth between $70 million and $431 million, depending on which multiplier you use. My source for these equivalencies was the Projects section of MeasuringWorth .com/m/calculators/uscompare, accessed January 20, 2015.

3 "Alaskan Way Viaduct Rims Seattle's Waterfront," *Seattle P-I*, April 5, 1953. The Battery Street Tunnel opened in 1954, after about a hundred thousand cubic yards of dirt had been removed. This dirt ended up as fill in Lake Union along Westlake Avenue and along Pier 89. Not until 1966 was the final off-ramp, at Seneca Street, completed.

4 In December 2013, after traveling a little more than a thousand feet, Bertha stopped working because of overheating problems with its internal bearing seals. As I write this, the party who is at fault has still not been determined. Did Seattle get a lemon when it purchased Bertha, or did the Seattle tunnel crew make an error in cutting through a steel pipe and then continue drilling when they shouldn't have?

5 Seattle Department of Transportation, *Elliott Bay Seawall Project: Final Environmental Impact Statement* (Seattle: City of Seattle, March 2013), ES-1.

6 The first quote is by Greg Nickels and is from Bob Young, "Nickels Presents Vision for Waterfront—Alaskan Way Tunnel a Key Element," *Seattle Times*, July 27, 2006. The second quote is by city council member Jean Godden, from Bill Lucia, "Seattle's Waterfront: Visions of Hottubs and Gardens, but Where's the Cash," *Crosscut*, March 6, 2014, http://crosscut.com/2014/03/06/politics-government/119041/waterfront-park-seattle-james-corner-cost/.

7 These are known scientifically as perigean spring tides. *Spring* refers to the idea that the tides will spring back to normal.

8 Conversation with James Rufo-Hill, March 18, 2014.

9 Conversation with Lara Whitely Binder, March 20, 2014.

10 One agency employee told me that it's motivating to live in a community where people support these ideas. In some places, planners aren't even allowed to discuss climate change.

11 From an unsigned email sent to the author by the Seawall Project Street Team, May 2, 2014.

12 Monica Huang, *Planning for Sea Level Rise: The Current State of Science Vulnerability of Port of Seattle Properties to Sea Level Rise, and Possible Adaptation Strategies* (Seattle: Port of Seattle, September 14, 2012), 43.

13 Robert E. Kayen and Walter A. Barnhardt, *Seismic Stability of the Duwamish River Delta, Seattle, Washington*, USGS Prof Paper 1661-E (Washington, DC: U.S. Department of the Interior, U.S. Geological Survey, 2007), 9.

14 James Gelling, email correspondence with the author, May 19, 2014.

Index

E

F

G

H

I

J

Q

R

S

T

U

V

W

Y

SEATTLE'S COMING RE